AN AGRICULTURAL TESTAMENT

by

Albert Howard

Albatross Publishers
Naples, Italy
2018

*Originally Published in London by Oxford
University Press, 1940*

ISBN 978-1-946963-18-5

©2018 *Albatross Publishers*

TO

GABRIELLE
WHO IS NO MORE

The Earth, that 's Nature's Mother, is her tomb;
What is her burying grave, that is her womb.

Romeo and Juliet.

And Nature, the old nurse, took
  The child upon her knee,
Saying: 'Here is a story-book
  Thy Father has written for thee.'

'Come, wander with me,' she said,
  'Into regions yet untrod;
And read what is still unread
  In the manuscripts of God.'

LONGFELLOW
The Fiftieth Birthday of Agassiz.

PREFACE

SINCE the Industrial Revolution the processes of growth have been speeded up to produce the food and raw materials needed by the population and the factory. Nothing effective has been done to replace the loss of fertility involved in this vast increase in crop and animal production. The consequences have been disastrous. Agriculture has become unbalanced: the land is in revolt: diseases of all kinds are on the increase: in many parts of the world Nature is removing the worn-out soil by means of erosion.

The purpose of this book is to draw attention to the destruction of the earth's capital—the soil; to indicate some of the consequences of this; and to suggest methods by which the lost fertility can be restored and maintained. This ambitious project is founded on the work and experience of forty years, mainly devoted to agricultural research in the West Indies, India, and Great Britain. It is the continuation of an earlier book—*The Waste Products of Agriculture*, published in 1931—in which the Indore method for maintaining soil fertility by the manufacture of humus from vegetable and animal wastes was described.

During the last nine years the Indore Process has been taken up at many centres all over the world. Much additional information on the role of humus in agriculture has been obtained. I have also had the leisure to bring under review the existing systems of farming as well as the organization and purpose of agricultural research. Some attention has also been paid to the bio-dynamic methods of agriculture in Holland and in Great Britain, but I remain unconvinced that the disciples of Rudolph Steiner can offer any real explanation of natural laws or have yet provided any practical examples which demonstrate the value of their theories.

The general results of all this are set out in this my Agricultural Testament. No attempt has been made to disguise the conclusions reached or to express them in the language of diplomacy. On the contrary, they have been stated with the utmost frankness. It is hoped that they will be discussed with the same freedom and that they will open up new lines of thought and eventually lead to effective action.

It would not have been possible to have written this book without the help and encouragement of a former colleague in India, Mr. George Clarke, C.I.E., who held the post of Director of Agriculture in the United Provinces for ten years (1921–31). He very generously placed at my disposal his private notes on the agriculture of the Provinces covering a period of over twenty years, and has discussed with me during the last three years practically everything in this book. He read many of the Chapters when they were first drafted, and made a number of suggestions which have been incorporated in the text.

Many who are engaged in practical agriculture all over the world and who have adopted the Indore Process have contributed to this book. In a few cases mention of this assistance has been made in the text. It is impossible to refer to all the correspondents who have furnished progress reports and have so freely reported their results. These provided an invaluable collection of facts and observations which has amply confirmed my own experience.

Great stress has been laid on a hitherto undiscovered factor in nutrition—the mycorrhizal association—the living fungous bridge between humus in the soil and the sap of plants. The existence of such a symbiosis was first suggested to me on reading an account of the remarkable results with conifers, obtained by Dr. M. C. Rayner at Wareham in Dorset in connexion with the operations of the Forestry Commission. If mycorrhiza occurs generally in the plantation industries and also in our crops, an explanation of such things as the development of quality, disease resistance, and the running out of the variety, as well as the slow deterioration of the soil which follows the use of artificial manures, would be provided. I accordingly took steps to collect a wide range of specimens likely to contain mycorrhiza, extending over the whole of tropical and temperate agriculture. I am indebted to Dr. Rayner and to Dr. Ida Levisohn for the detailed examination of this material. They have furnished me with many valuable and suggestive technical reports. For the interpretation of these laboratory results, as set out in the following pages, I am myself solely responsible.

I am indebted to a number of Societies for permission to reproduce information and illustrations which have already been published. Two other organizations have allowed me to incorporate results which might well have been regarded as confidential. The Royal Society of London has permitted me to reprint, in the Chapter on Soil Aeration, a précis of an illustrated paper which appeared in their *Proceedings*. The Royal Society of Arts has provided the blocks for the section on sisal waste. The Royal Sanitary Institute has agreed to the reproduction in full of a paper read at the Health Congress, held at Portsmouth in July 1938. The *British Medical Journal* has placed at my disposal the information contained in an article by Dr. Lionel J. Picton, O.B.E. The publishers of Dr. Waksman's monograph on *Humus* have allowed me to reprint two long extracts relating to the properties of humus. Messrs. Arthur Guinness, Sons & Co., Limited, have agreed to the publication of the details of the composting of town wastes in their hop garden at Bodiam. Messrs. Walter Duncan & Co. have allowed the Manager of the Gandrapara Tea Garden to contribute an illustrated article on the composting of wastes on this fine estate. Captain J. M. Moubray has sent me a very interesting summary of the work he is doing at Chipoli in Southern Rhodesia, which is given in Appendix B.

In making the Indore Process widely known, a number of journals have rendered yeoman service. In Great Britain *The Times* and the *Journal of the Royal Society of Arts* have published a regular series of letters and articles. In South Africa the *Farmer's Weekly* has from the beginning urged the agricultural community to increase the humus content of the soil. In Latin America the planters owe much to the *Revista del Instituto de Defensa del Café de Costa Rica*.

Certain of the largest tea companies in London, Messrs. James Finlay & Co., Walter Duncan & Co., the Ceylon Tea Plantations Company, Messrs. Octavius Steel & Co., and others, most generously made themselves responsible over a period of two years for a large part of the office expenses connected with the working out and application to the plantation industries of the Indore Process. They also defrayed the expenses of a tour to the tea estates of India and Ceylon in 1937–8. These arrangements were very kindly made on my behalf by Mr. G. H. Masefield, Chairman of the Ceylon Tea Plantations Company.

In the work of reducing to order the vast mass of correspondence and notes on soil fertility, which have accumulated, and in getting the book into its final shape, I owe much to the ability and devotion of my private secretary, Mrs. V. M. Hamilton.

A. H.

BLACKHEATH,
1 *January* 1940

CONTENTS

I. INTRODUCTION 1
    NATURE'S METHODS OF SOIL MANAGEMENT . . . 1
    THE AGRICULTURE OF THE NATIONS WHICH HAVE PASSED
    AWAY 5
    THE PRACTICES OF THE ORIENT 9
    THE AGRICULTURAL METHODS OF THE OCCIDENT . . 17

PART I

THE PART PLAYED BY SOIL FERTILITY IN AGRICULTURE

II. THE NATURE OF SOIL FERTILITY . . . 22

III. THE RESTORATION OF FERTILITY . . . 32

PART II

THE INDORE PROCESS

IV. THE INDORE PROCESS 39
    THE RAW MATERIALS NEEDED 41
    PITS VERSUS HEAPS 45
    CHARGING THE HEAPS OR PITS 46
    TURNING THE COMPOST 48
    THE STORAGE OF HUMUS 50
    OUTPUT 50

V. PRACTICAL APPLICATIONS OF THE INDORE
    PROCESS 53
    COFFEE 53
    TEA 56
    SUGAR-CANE 66
    COTTON 71
    SISAL 75
    MAIZE 78
    RICE 80
    VEGETABLES 82
    VINE 85

VI. DEVELOPMENTS OF THE INDORE PROCESS . 87
    GREEN-MANURING 87
    THE SAFEGUARDING OF NITRATE ACCUMULATIONS . 91
    THE PRODUCTION OF HUMUS 93
    THE SAFEGUARDING OF NITRATES FOLLOWED BY THE MANU-
    FACTURE OF HUMUS 94
    THE REFORM OF GREEN-MANURING . . . 95

VII. DEVELOPMENTS OF THE INDORE PROCESS, *Cont.*

    GRASS-LAND MANAGEMENT   96

VIII. DEVELOPMENTS OF THE INDORE PROCESS, *Cont.*

    THE UTILIZATION OF TOWN WASTES   .   .   .   104

PART III

HEALTH, INDISPOSITION, AND DISEASE IN AGRICULTURE

IX. SOIL AERATION   116

    THE SOIL AERATION FACTOR IN RELATION TO GRASS AND

        TREES   117

        The Root System of Deciduous Trees   .   .   .   120

        The Root System of Evergreens   124

        The Harmful Effect of Grass   125

        The Effect of Aeration Trenches on Young Trees Under

        Grass   128

        The Cause of the Harmful Effect of Grass   .   .   130

        Forest Trees and Grass   132

    THE AERATION OF THE SUB-SOIL   .   .   .   136

X. SOME DISEASES OF THE SOIL   .   .   .   140

    SOIL EROSION   140

    THE FORMATION OF ALKALI LAND   .   .   .   147

XI. THE RETREAT OF THE CROP AND THE ANIMAL

    BEFORE THE PARASITE   156

    HUMUS AND DISEASE RESISTANCE   .   .   .   165

    THE MYCORRHIZAL ASSOCIATION AND DISEASE   .   .   166

    THE INVESTIGATIONS OF TO-MORROW   .   .   .   168

XII. SOIL FERTILITY AND NATIONAL HEALTH .   .   171

PART IV

AGRICULTURAL RESEARCH

XIII. A CRITICISM OF PRESENT-DAY AGRICULTURAL

    RESEARCH   181

XIV. A SUCCESSFUL EXAMPLE OF AGRICULTURAL

    RESEARCH   200

PART V

CONCLUSIONS AND SUGGESTIONS

XV. A FINAL SURVEY   219

APPENDIXES

A. Compost Manufacture on a Tea Estate in Bengal.   .   225

B. Compost Making at Chipoli, Southern Rhodesia .   .   229

C. The Manufacture of Humus from the Wastes of the Town
   and the Village   235

INDEX   243

LIST OF ILLUSTRATIONS

PLATES

I. Conversion of Sisal Waste . . . *Facing page* 76

II. Conversion of Sisal Waste 78

III. Rainfall, Temperature, Humidity, and Drainage, Pusa, 1922 . 116

IV. The Harmful Effect of Grass on Fruit Trees, Pusa, 1923 . 120

V. Plum (*Prunus communis*, Huds.) 128

VI. Guava (*Psidium Guyava*, L.) 132

VII. Nitrate Accumulation in the Gangetic Alluvium . . 208

VIII. Nitrate Accumulation, Green-manure Experiment, Shahjahanpur, 1928–9 212

IX. Green-manure Experiment, Shahjahanpur, 1928–9 . . 216

X. Plan of the Compost Factory, Gandrapara Tea Estate . 226

XI. Composting at Gandrapara 226

XII. Composting at Gandrapara 228

XIII. Compost-making at Chipoli, Southern Rhodesia . . 232

FIGURES

1. A model lay-out for 20 cottages 114

2. Plan of experimental fruit area, Pusa 119

3. Hot weather and monsoon foliage of the custard apple . . 123

4. The effect of burrowing rats on the growth of the plum under
grass 128

5. Carbon dioxide in soil atmosphere, Pusa, 1919 . . . 132

6. Trench system at Shahjahanpur 205

7. Earthing up sugar-cane at Shahjahanpur, July 10th, 1919 . 206

8. Plan and working details of composting pits at Tollygunge,
Calcutta 236

9. Long-handled drag-rake and fork used in composting . . 238

10. Plan of compost factory at Tollygunge, Calcutta, after one month's
use 239

11. Plan of a simple composting trench for a village . . . 241

CHAPTER I

INTRODUCTION

THE maintenance of the fertility of the soil is the first condition of any permanent system of agriculture. In the ordinary processes of crop production fertility is steadily lost: its continuous restoration by means of manuring and soil management is therefore imperative.

In the study of soil fertility the first step is to bring under review the various systems of agriculture which so far have been evolved. These fall into four main groups: (1) the methods of Nature—the supreme farmer—as seen in the primeval forest, in the prairie, and in the ocean; (2) the agriculture of the nations which have passed away; (3) the practices of the Orient, which have been almost unaffected by Western science; and (4) the methods in vogue in regions like Europe and North America to which a large amount of scientific attention has been paid during the last hundred years.

NATURE'S METHODS OF SOIL MANAGEMENT

Little or no consideration is paid in the literature of agriculture to the means by which Nature manages land and conducts her water culture. Nevertheless, these natural methods of soil management must form the basis of all our studies of soil fertility.

What are the main principles underlying Nature's agriculture? These can most easily be seen in operation in our woods and forests.

Mixed farming is the rule: plants are always found with animals: many species of plants and of animals all live together. In the forest every form of animal life, from mammals to the simplest invertebrates, occurs. The vegetable kingdom exhibits a similar range: there is never any attempt at monoculture: mixed crops and mixed farming are the rule.

The soil is always protected from the direct action of sun, rain, and wind. In this care of the soil strict economy is the watchword: nothing is lost. The whole of the energy of sunlight is made use of by the foliage of the forest canopy and of the undergrowth. The leaves also break up the rainfall into fine spray so that it can the

more easily be dealt with by the litter of plant and animal remains
which provide the last line of defence of the precious soil. These
methods of protection, so effective in dealing with sun and rain,
also reduce the power of the strongest winds to a gentle air current.

The rainfall in particular is carefully conserved. A large portion
is retained in the surface soil: the excess is gently transferred to the
subsoil and in due course to the streams and rivers. The fine
spray created by the foliage is transformed by the protective
ground litter into thin films of water which move slowly downwards, first into the humus layer and then into the soil and sub-
soil. These latter have been made porous in two ways: by the
creation of a well-marked crumb structure and by a network of
drainage and aeration channels made by earthworms and other
burrowing animals. The pore space of the forest soil is at its maximum so that there is a large internal soil surface over which the
thin films of water can creep. There is also ample humus for the
direct absorption of moisture. The excess drains away slowly by
way of the subsoil. There is remarkably little run-off, even from
the primeval rain forest. When this occurs it is practically clear
water. Hardly any soil is removed. Nothing in the nature of soil
erosion occurs. The streams and rivers in forest areas are always
perennial because of the vast quantity of water in slow transit
between the rainstorms and the sea. There is therefore little or no
drought in forest areas because so much of the rainfall is retained
exactly where it is needed. There is no waste anywhere.

The forest manures itself. It makes its own humus and supplies
itself with minerals. If we watch a piece of woodland we find that
a gentle accumulation of mixed vegetable and animal residues is
constantly taking place on the ground and that these wastes are
being converted by fungi and bacteria into humus. The processes
involved in the early stages of this transformation depend throughout on oxidation: afterwards they take place in the absence of air.
They are sanitary. There is no nuisance of any kind—no smell,
no flies, no dustbins, no incinerators, no artificial sewage system,
no water-borne diseases, no town councils, and no rates. On the
contrary, the forest affords a place for the ideal summer holiday:
sufficient shade and an abundance of pure fresh air. Nevertheless,
all over the surface of the woods the conversion of vegetable and

animal wastes into humus is never so rapid and so intense as during the holiday months—July to September.

The mineral matter needed by the trees and the undergrowth is obtained from the subsoil. This is collected in dilute solution in water by the deeper roots, which also help in anchoring the trees. The details of root distribution and the manner in which the subsoil is thoroughly combed for minerals are referred to in a future chapter (p. 135). Even in soils markedly deficient in phosphorus trees have no difficulty in obtaining ample supplies of this element. Potash, phosphate, and other minerals are always collected *in situ* and carried by the transpiration current for use in the green leaves. Afterwards they are either used in growth or deposited on the floor of the forest in the form of vegetable waste—one of the constituents needed in the synthesis of humus. This humus is again utilized by the roots of the trees. Nature's farming, as seen in the forest, is characterized by two things: (1) a constant circulation of the mineral matter absorbed by the trees; (2) a constant addition of new mineral matter from the vast reserves held in the subsoil. There is therefore no need to add phosphates: there is no necessity for more potash salts. No mineral deficiencies of any kind occur. The supply of all the manure needed is automatic and is provided either by humus or by the soil. There is a natural division of the subject into organic and inorganic. Humus provides the organic manure: the soil the mineral matter.

The soil always carries a large fertility reserve. There is no hand to mouth existence about Nature's farming. The reserves are carried in the upper layers of the soil in the form of humus. Yet any useless accumulation of humus is avoided because it is automatically mingled with the upper soil by the activities of burrowing animals such as earthworms and insects. The extent of this enormous reserve is only realized when the trees are cut down and the virgin land is used for agriculture. When plants like tea, coffee, rubber, and bananas are grown on recently cleared land, good crops can be raised without manure for ten years or more. Like all good administrators, therefore, Nature carries strong liquid reserves effectively invested. There is no squandering of these reserves to be seen anywhere.

The crops and live stock look after themselves. Nature has

never found it necessary to design the equivalent of the spraying machine and the poison spray for the control of insect and fungous pests. There is nothing in the nature of vaccines and serums for the protection of the live stock. It is true that all kinds of diseases are to be found here and there among the plants and animals of the forest, but these never assume large proportions. The principle followed is that the plants and animals can very well protect themselves even when such things as parasites are to be found in their midst. Nature's rule in these matters is to live and let live.

If we study the prairie and the ocean we find that similar principles are followed. The grass carpet deals with the rainfall very much as the forest does. There is little or no soil erosion: the run-off is practically clear water. Humus is again stored in the upper soil. The best of the grassland areas of North America carried a mixed herbage which maintained vast herds of bison. No veterinary service was in existence for keeping these animals alive. When brought into cultivation by the early settlers, so great was the store of fertility that these prairie soils yielded heavy crops of wheat for many years without live stock and without manure.

In lakes, rivers, and the sea mixed farming is again the rule: a great variety of plants and animals are found living together: nowhere does one find monoculture. The vegetable and animal wastes are again dealt with by effective methods. Nothing is wasted. Humus again plays an important part and is found everywhere in solution, in suspension, and in the deposits of mud. The sea, like the forest and the prairie, manures itself.

The main characteristic of Nature's farming can therefore be summed up in a few words. Mother earth never attempts to farm without live stock; she always raises mixed crops; great pains are taken to preserve the soil and to prevent erosion; the mixed vegetable and animal wastes are converted into humus; there is no waste; the processes of growth and the processes of decay balance one another; ample provision is made to maintain large reserves of fertility; the greatest care is taken to store the rainfall; both plants and animals are left to protect themselves against disease.

In considering the various man-made systems of agriculture, which so far have been devised, it will be interesting to see how far

Nature's principles have been adopted, whether they have ever been improved upon, and what happens when they are disregarded.

THE AGRICULTURE OF THE NATIONS WHICH HAVE PASSED AWAY

The difficulties inherent in the study of the agriculture of the nations which are no more are obvious. Unlike their buildings, where it is possible from a critical study of the buried remains of cities to reproduce a picture of bygone civilizations, the fields of the ancients have seldom been maintained. The land has either gone back to forest or has been used for one system of farming after another.

In one case, however, the actual fields of a bygone people have been preserved together with the irrigation methods by which these lands were made productive. No written records, alas, have come down to us of the staircase cultivation of the ancient Peruvians, perhaps one of the oldest forms of Stone Age agriculture. This arose either in mountains or in the upland areas under grass because of the difficulty, before the discovery of iron, of removing the dense forest growth. In Peru irrigated staircase farming seems to have reached its highest known development. More than twenty years ago the National Geographical Society of the United States sent an expedition to study the relics of this ancient method of agriculture, an account of which was published by O. F. Cook in the Society's Magazine of May 1916, under the title: 'Staircase Farms of the Ancients.' The system of the megalithic people of old Peru was to construct a stairway of terraced fields up the slopes of the mountains, tier upon tier, sometimes as many as fifty in number. The outer retaining walls of these terraces were made of large stones which fit into one another with such accuracy that even at the present day, like those of the Egyptian pyramids, a knife blade cannot be inserted between them. After the retaining wall was built, the foundation of the future field was prepared by means of coarse stones covered with clay. On this basis layers of soil, several feet thick, originally imported from beyond the great mountains, were super-imposed and then levelled for irrigation. The final result was a small flat field with only just sufficient slope for artificial watering. In other words, a series of huge flower pots, each provided with ample drainage below, was prepared with incredible labour by this ancient people for their crops. Such were the megalithic achievements in agriculture, beside which 'our undertakings sink into insignificance in face of what this vanished race accomplished. The narrow floors and steep walls of rocky valleys that would appear utterly worthless and hopeless to our engineers were transformed, literally made over, into fertile lands and were the homes of teeming populations in pre-historic days' (O. F. Cook). The engineers of old Peru did what they did through necessity because iron, steel, reinforced concrete, and the modern power units had not been invented. The plunder of the forest soil was beyond their reach.

These terraced fields had to be irrigated. Water had to be led to them over immense distances by means of aqueducts. Prescott states that one which traversed the district of Condesuyu measured between four and five hundred miles. Cook gives a photograph of one of these channels as a thin dark line traversing a steep mountain wall many hundreds of feet above the valley.

These ancient methods of agriculture are represented at the present day by the terraced cultivation of the Himalayas, of the mountainous areas of China and Japan, and of the irrigated rice fields so common in the hills of South India, Ceylon, and the Malayan Archipelago. Conway's description, published in 1894, of the terraces of Hunza on the North-West Frontier of India and of the canal, carried for long distances across the face of precipices to the one available supply of perennial water—the torrent from the Ultor glacier—tallies almost completely with what he found in 1901 in the Bolivian Andes. This distinguished scholar and mountaineer considered that the native population of Hunza of the present day is living in a stage of civilization that must bear no little likeness to that of the Peruvians under Inca government. An example of this ancient method of farming has thus been preserved through the ages. In a future chapter (p. 173) the relation which exists between the nutritional value of the food grown on these irrigated terraces and the health of the people will be discussed. This relic of the past is interesting from the point of view of quality in food as well as from its historical value.

Some other systems of agriculture of the past have come down to us in the form of written records which have furnished ample material for constructive research. In the case of Rome in particular a fairly complete account of the position of agriculture, from the period of the monarchy to the fall of the Roman Empire, is available; the facts can be conveniently followed in the writings of Mommsen, Heitland, and other scholars. In the case of Rome the Servian Reform (Servius Tullius, 578–534 B.C.) shows very clearly not only that the agricultural class originally preponderated in the State but also that an effort was made to maintain the collective body of freeholders as the pith and marrow of the community. The conception that the constitution itself rested on the freehold system permeated the whole policy of Roman war and conquest. The aim of war was to increase the number of its freehold members.

'The vanquished community was either compelled to merge entirely into the yeomanry of Rome, or, if not reduced to this extremity, it was required, not to pay a war contribution or a fixed tribute, but to cede a portion, usually a third part, of its domain, which was thereupon regularly occupied by Roman farms. Many nations have gained victories and made conquests as the Romans did; but none has equalled the Roman in thus making the ground he had won his own by the sweat of his brow, and in securing by the ploughshare what had been gained by the lance. That which is gained by war may be wrested from the grasp by war again, but it is not so with the conquests made by the plough; whilst the Romans lost many battles, they scarcely ever on making peace ceded Roman soil, and for this result they were indebted to the tenacity with which the farmers clung to their fields and homesteads. The strength of man and of the State lies in their dominion over the soil; the strength of Rome was built on the most extensive and immediate mastery of her citizens over the soil, and on the compact unity of the body which thus acquired so firm a hold.' (Mommsen.)

These splendid ideals did not persist. During the period which elapsed between the union of Italy and the subjugation of Carthage, a gradual decay of the farmers set in; the small-holdings ceased to yield any substantial clear return; the cultivators one by one faced ruin; the moral tone and frugal habits of the earlier ages of the Republic were lost; the land of the Italian farmers became merged into the larger estates. The landlord capitalist became the centre of the subject. He not only produced at a cheaper rate than the farmer because he had more land, but he began to use slaves. The same space which in the olden time, when small-holdings prevailed, had supported from a hundred to a hundred and fifty families was now occupied by one family of free persons and about fifty, for the most part unmarried, slaves. 'If this was the remedy by which the decaying national economy was to be restored to vigour, it bore, unhappily, an aspect of extreme resemblance to disease' (Mommsen). The main causes of this decline appear to have been fourfold: the constant drain on the manhood of the country-side by the legions, which culminated in the two long wars with Carthage; the operations of the Roman capitalist land-lords which 'contributed quite as much as Hamilcar and Hannibal to the decline in the vigour and the number of the Italian people' (Mommsen); failure to work out a balanced agriculture between crops and live stock and to maintain the fertility of the soil; the employment of slaves instead of free labourers. During this period the wholesale commerce of Latium passed into the hands of the large landed proprietors who at the same time were the speculators and capitalists. The natural consequence was the destruction of the middle classes, particularly of the small-holders, and the development of landed and moneyed lords on the one hand and of an agricultural proletariat on the other. The power of capital was greatly enhanced by the growth of the class of tax-farmers and contractors to whom the State farmed out its indirect revenues for a fixed sum. Subsequent political and social conflicts did not give real relief to the agricultural community. Colonies founded to secure Roman sovereignty over Italy provided farms for the agricultural proletariat, but the root causes of the decline in agriculture were not removed in spite of the efforts of Cato and other reformers. A capitalist system of which the apparent inter-ests were fundamentally opposed to a sound agriculture remained supreme. The last half of the second century saw degradation and more and more decadence. Then came Tiberius Gracchus and the Agrarian Law with the appointment of an official commission to counteract the diminution of the farmer class by the compre-hensive establishment of new small-holdings from the whole Italian landed property at the disposal of the State: eighty thousand new Italian farmers were provided with land. These efforts to restore agriculture to its rightful place in the State were accompanied by many improvements in Roman agriculture which, unfortunately, were most suitable for large estates. Land no longer able to produce corn became pasture; cattle now roamed over large ranches; the vine and the olive were cultivated with commercial success. These systems of agriculture, however, had to be carried on with slave labour, the supply of which had to be maintained by constant importation. Such extensive methods of farming naturally failed to supply sufficient food for the population of Italy. Other countries were called upon to furnish essential foodstuffs; province after province was conquered to feed the growing proletariat with corn. These areas in turn slowly yielded to the same decline which had taken place in Italy. Finally the wealthy classes abandoned the depopulated remnants of the mother country and built themselves a new capital at Constantinople. The situation had to be saved by a migration to fresh lands. In their new capital the Romans relied on the unexhausted fertility of Egypt as well as on that of Asia Minor and the Balkan and Danubian provinces.

Judged by the ordinary standards of achievement the agricultural history of the Roman Empire ended in failure due to inability to realize the fundamental principle that the maintenance of soil fertility coupled with the legitimate claims of the agricultural population should never have been allowed to come in conflict with the operations of the capitalist. The most important possession of a country is its population. If this is maintained in health and vigour everything else will follow; if this is allowed to decline nothing, not even great riches, can save the country from eventual ruin. It follows, therefore, that the strongest possible support of capital must always be a prosperous and contented country-side. A working compromise between agriculture and finance should therefore have been evolved. Failure to achieve this naturally ended in the ruin of both.

THE PRACTICES OF THE ORIENT

In the agriculture of Asia we find ourselves confronted with a system of peasant farming which in essentials soon became stabilized. What is happening to-day in the small fields of India and China took place many centuries ago. There is here no need to study historical records or to pay a visit to the remains of the megalithic farming of the Andes. The agricultural practices of the Orient have passed the supreme test—they are almost as permanent as those of the primeval forest, of the prairie or of the ocean. The small-holdings of China, for example, are still maintaining a steady output and there is no loss of fertility after forty centuries of management. What are the chief characteristics of this Eastern farming?

The holdings are minute. Taking India as an example, the relation between man power and cultivated area is referred to in the Census Report of 1931 as follows: 'For every agriculturalist there is 2·9 acres of cropped land of which 0·65 of an acre is irrigated. The corresponding figures of 1921 are 2·7 and 0·61.' These figures illustrate how intense is the struggle for existence in this portion of the tropics. These small-holdings are often cultivated by extensive methods (those suitable for large areas) which utilize neither the full energies of man or beast nor the potential fertility of the soil.

If we turn to the Far East, to China and Japan, a similar system of small-holdings is accompanied by an even more intense pressure of population both human and bovine. In the introduction to *Farmers of Forty Centuries,* King states that the three main islands of Japan had in 1907 a population of 46,977,000, maintained on 20,000 square miles of cultivated fields. This is at the rate of 2,349 to the square mile or more than three people to each acre. In addition, Japan fed on each square mile of cultivation a very large animal population—69 horses and 56 cattle, nearly all employed in labour; 825 poultry; 13 swine, goats, and sheep. Although no accurate statistics are available in China, the examples quoted by King reveal a condition of affairs not unlike that in Japan. In the Shantung Province a farmer with a family of twelve kept one donkey, one cow, and two pigs on 2·5 acres of cultivated land— a density of population at the rate of 3,072 people, 256 donkeys, 256 cattle, and 512 pigs per square mile. The average of seven Chinese holdings visited gave a maintenance capacity of 1,783 people, 212 cattle or donkeys, and 399 pigs—nearly 2,000 consumers and 400 rough food transformers per square mile of farmed land. In comparison with these remarkable figures, the corresponding statistics for 1900 in the case of the United States per square mile were: population 61, horses and mules 30.

Food and forage crops are predominant. The primary function

of Eastern agriculture is to supply the cultivators and their cattle with food. This automatically follows because of the pressure of the population on the land: the main hunger the soil has to appease is that of the stomach. A subsidiary hunger is that of the machine which needs raw materials for manufacture. This extra hunger is new but has developed considerably since the opening of the Suez Canal in 1869 (by which the small fields of the cultivator have been brought into effective contact with the markets of the West) and the establishment of local industries like cotton and jute. To both these hungers soil fertility has to respond. We know from long experience that the fields of India can respond to the hunger of the stomach. Whether they can fulfil the added demands of the machine remains to be seen. The Suez Canal has only been in operation for seventy years. The first cotton mill in India was opened in 1818 at Fort Gloster, near Calcutta. The jute industry of Bengal has grown up within a century. Jute was first exported in 1838. The first jute mill on the Hoogly began operations in 1855. These local industries as well as the export trade in raw products for the use of the factories of the West are an extra drain on soil fertility. Their future well-being and indeed their very existence is only possible provided adequate steps are taken to maintain this fertility. There is obviously no point in establishing cotton and jute mills in India, in founding trading agencies like those of Calcutta and in building ships for the conveyance of raw products unless such enterprises are stable and permanent. It would be folly and an obvious waste of capital to pursue such activities if they are founded only on the existing store of soil fertility. All concerned in the hunger of the machine— government, financiers, manufacturers, and distributors—must see to it that the fields of India are equal to the new burden which has been thrust upon her during the last fifty years or so. The demands of commerce and industry on the one hand and the fertility of the soil on the other must be maintained in correct relation the one to the other.

The response of India to the two hungers—the stomach and the machine—will be evident from a study of Table 1, in which the area in acres under food and fodder crops is compared with that under money crops.

The chief food crops in order of importance are rice, pulses, millets, wheat, and fodder crops. The money crops are more varied; cotton and oil seeds are the most important, followed by jute and other fibres, tobacco, tea, coffee, and opium. It will be

TABLE I

Agricultural Statistics of British India, 1935–6

Area, in acres, under food and fodder crops	
Rice	79,888,000
Millets	38,144,000
Wheat	25,150,000
Gram	14,897,000
Pulses and other food grains	29,792,000
Fodder crops	10,791,000
Condiments, spices, fruits, vegetables, and miscellaneous food crops	8,308,000
Barley	6,178,000
Maize	6,211,000
Sugar	4,038,000
Total, food and fodder crops	223,397,000

Area, in acres, under money crops	
Cotton	15,761,000
Oil seeds, chiefly ground-nuts, sesamum, rape, mustard, and linseed	15,662,000
Jute and other fibres	2,706,000
Dyes, tanning materials, drugs, narcotics, and miscellaneous money crops	1,458,000
Tobacco	1,230,000
Tea	787,000
Coffee	97,000
Indigo	40,000
Opium	10,000
Total, money crops	37,751,000

seen that food and fodder crops comprise 86 per cent. of the total area under crops and that money crops, as far as extent is concerned, are less important, and constitute only one-seventh of the total cultivated area.

One interesting change in the production of Indian food crops has taken place during the last twenty-five years. The output of sugar used to be insufficient for the towns, and large quantities were imported from Java, Mauritius, and the continent of Europe. To-day, thanks to the work at Shahjahanpur in the United Provinces, the new varieties of cane bred at Coimbatore and the protection now enjoyed by the sugar industry, India is almost self-supporting as far as sugar is concerned. The pre-war average amount of sugar imported was 634,000 tons; in 1937–8 the total had fallen to 14,000 tons.

Mixed crops are the rule. In this respect the cultivators of the Orient have followed Nature's method as seen in the primeval forest. Mixed cropping is perhaps most universal when the cereal crop is the main constituent. Crops like millets, wheat, barley, and maize are mixed with an appropriate subsidiary pulse, sometimes a species that ripens much later than the cereal. The pigeon pea (*Cajanus indicus* Spreng.), perhaps the most important leguminous crop of the Gangetic alluvium, is grown either with millets or with maize. The mixing of cereals and pulses appears to help both crops. When the two grow together the character of the growth improves. Do the roots of these crops excrete materials useful to each other? Is the mycorrhizal association found in the roots of these tropical legumes and cereals the agent involved in this excretion? Science at the moment is unable to answer these questions: she is only now beginning to investigate them. Here we have another instance where the peasants of the East have anticipated and acted upon the solution of one of the problems which Western science is only just beginning to recognize. Whatever may be the reason why crops thrive best when associated in suitable combinations, the fact remains that mixtures generally give better results than monoculture. This is seen in Great Britain in the growth of dredge corn, in mixed crops of wheat and beans, vetches and rye, clover and rye-grass, and in intensive vegetable growing under glass. The produce raised under Dutch lights has noticeably increased since the mixed cropping of the Chinese vegetable growers of Australia has been copied.[1]

A balance between live stock and crops is always maintained. Although crops are generally more important than animals in Eastern agriculture, we seldom or never find crops without animals. This is because oxen are required for cultivation and buffaloes for

[1] Mr. F. A. Secrett was, I believe, the first to introduce this system on a large scale into Great Britain. He informed me that he saw it for the first time at Melbourne.

milk.[1] Nevertheless, the waste products of the animal, as is often the case in other parts of the world, are not always fully utilized for the land. The Chinese have for ages past recognized the importance of the urine of animals and the great value of animal wastes in the preparation of composts. In India far less attention is paid to these wastes and a large portion of the cattle dung available is burnt for fuel. On the other hand, in most Oriental countries human wastes find their way back to the land. In China these are collected for manuring the crops direct. In India they are concentrated on the zone of highly manured land immediately round each village. If the population or a portion of it could be persuaded to use a more distant zone for a few years, the area of village lands under intensive agriculture could at least be doubled. Here is an opportunity for the new system of government in India to raise production without the expenditure of a single rupee. In India there are 500,000 villages each of which is surrounded by a zone of very fertile land which is constantly being over-manured by the habits of the people. If we examine the crops grown on this land we find that the yields are high and the plants are remarkably free from disease. Although half a million examples of the connexion between a fertile soil and a healthy plant exist in India alone, and these natural experiments have been in operation for centuries before experiment stations like Rothamsted were ever thought of, modern agricultural science takes no notice of the results and resolutely refuses to accept them as evidence, largely because they lack the support furnished by the higher mathematics. They also dispose of one of the ideas of the disciples of Rudolph Steiner, who argue that the use of human wastes in agriculture is harmful.

Leguminous plants are common. Although it was not till 1888, after a protracted controversy lasting thirty years, that Western science finally accepted as proved the important part played by pulse crops in enriching the soil, centuries of experience had

[1] The buffalo is the milch cow of the Orient and is capable not only of useful labour in the cultivation of rice, but also of living and producing large quantities of rich milk on a diet on which the best dairy cows of Europe and America would starve. The acclimatization of the Indian buffalo in the villages of the Tropics—Africa, Central America, the West Indies in particular—would do much to improve the fertility of the soil and the nutrition of the people.

taught the peasants of the East the same lesson. The leguminous crop in the rotation is everywhere one of their old fixed practices. In some areas, such as the Indo-Gangetic plain, one of these pulses—the pigeon pea—is also made use of as a subsoil cultivator. The deep spreading root system is used to promote the aeration of the closely packed silt soils, which so closely resemble those of the Holland Division of Lincolnshire in Great Britain.

Cultivation is generally superficial and is carried out by wooden ploughs furnished with an iron point. Soil-inverting ploughs, as used in the West for the destruction of weeds, have never been designed by Eastern peoples. The reasons for this appear to be two: (1) soil inversion for the destruction of weeds is not necessary in a hot climate where the same work is done by the sun for nothing; (2) the preservation of the level of the fields is essential for surface drainage, for preventing local waterlogging, and for irrigation. Another reason for this surface cultivation has recently been pointed out (p. 210). The store of nitrogen in the soil in the form of organic matter has to be carefully conserved: it is part of the cultivator's working capital. Too much cultivation and deep ploughing would oxidize this reserve and the balance of soil fertility would soon be destroyed.

Rice is grown whenever possible. By far the most important crop in the East is rice. In India, as has already been pointed out, the production of rice exceeds that of any two food crops put together. Whenever the soil and water supply permit, rice is invariably grown. A study of this crop is illuminating. At first sight rice appears to contradict one of the great principles of the agricultural science of the Occident, namely, the dependence of cereals on nitrogenous manures. Large crops of rice are produced in many parts of India on the same land year after year without the addition of any manure whatever. The rice fields of the country export paddy in large quantities to the centres of population or abroad, but there is no corresponding import of combined nitrogen.[1] Where does the rice crop obtain its nitrogen? One

[1] Taking Burma as an example of an area exporting rice beyond seas, during the twenty years ending 1924, about 25,000,000 tons of paddy have been exported from a tract roughly 10,000,000 acres in area. As unhusked rice contains about 1·2 per cent. of nitrogen the amount of this element, shipped overseas during twenty years or destroyed in the burning of the husk, is in the neighbourhood of 300,000 tons. As

source in all probability is fixation from the atmosphere in the submerged algal film on the surface of the mud. Another is the rice nursery itself, where the seedlings are raised on land heavily manured with cattle dung. Large quantities of nitrogen and other nutrients are stored in the seedling itself; this at transplanting time contains a veritable arsenal of reserves of all kinds which carry the plant successfully through this process and probably also furnish some of the nitrogen needed during subsequent growth. The manuring of the rice seedling illustrates a very general principle in agriculture, namely, the importance of starting a crop in a really fertile soil and so arranging matters that the plant can absorb a great deal of what it needs as early as possible in its development.

There is an adequate supply of labour. Labour is everywhere abundant, as would naturally follow from the great density of the rural population. Indeed, in India it is so great that if the leisure time of the cultivators and their cattle for a single year could be calculated as money at the local rates a perfectly colossal figure would be obtained. This leisure, however, is not altogether wasted. It enables the cultivators and their oxen to recover from the periods of intensive work which precede the sowing of the crops and which are needed at harvest time. At these periods time is everything: everybody works from sunrise to sunset. The preparation of the land and the sowing of the crops need the greatest care and skill; the work must be completed in a very short time so that a large labour force is essential.

It will be observed that in this peasant agriculture the great pressure of population on the soil results in poverty, most marked where, as in India, extensive methods are used on small-holdings which really need intensive farming. It is amazing that in spite of this unfavourable factor soil fertility should have been preserved for centuries: this is because natural means have been used and not artificial manures. The crops are able to withstand the

this constant drain of nitrogen is not made up for by the import of manure, we should expect to find a gradual loss of fertility. Nevertheless, this does not take place either in Burma or in Bengal, where rice has been grown on the same land year after year for centuries. Clearly the soil must obtain fresh supplies of nitrogen from somewhere, otherwise the crop would cease to grow. The only likely source is fixation from the atmosphere, probably in the submerged algal film on the surface of the mud. This is one of the problems of tropical agriculture which is now being investigated.

inroads of insects and fungi without a thin film of protective poison.

THE AGRICULTURAL METHODS OF THE OCCIDENT

If we take a wide survey of the contribution which is being made by the fields of the West, we find that they are engaged in trying to satisfy no less than three hungers: (1) the local hunger of the rural population, including the live stock; (2) the hunger of the growing urban areas, the population of which is unproductive from the point of view of soil fertility; and (3) the hunger of the machine avid for a constant stream of the raw materials required for manufacture. The urban population during the last century has grown out of all knowledge; the needs of the machine increase as it becomes more and more efficient; falling profits are met by increasing the output of manufactured articles. All this adds to the burden on the land and to the calls on its fertility. It will not be without interest to analyse critically the agriculture of the West and see how it is fitting itself for its growing task. This can be done by examining its main characteristics. These are as follows:

The holding tends to increase in size. There is a great variation in the size of the agricultural holdings of the West from the small family units of France and Switzerland to the immense collective farms of Russia and the spacious ranches of the United States and Argentina. Side by side with this growth in the size of the farm is the diminution of the number of men per square mile. In Canada, for example, the number of workers per 1,000 acres of cropped land fell from 26 in 1911 to 16 in 1926. Since these data were published the size of the working population has shrunk still further. This state of things has arisen from the scarcity and dearness of labour which has naturally led to the study of labour-saving devices.

Monoculture is the rule. Almost everywhere crops are grown in pure culture. Except in temporary leys, mixed crops are rare. On the rich prairie lands of North America even rotations are unknown: crops of wheat follow one another and no attempt is made to convert the straw into humus by means of the urine and dung of cattle. The straw is a tiresome encumbrance and is burnt off annually.

The machine is rapidly replacing the animal. Increasing mechanization is one of the main features of Western agriculture. Whenever a machine can be invented which saves human or animal labour its spread is rapid. Engines and motors of various kinds are the rule everywhere. The electrification of agriculture is beginning. The inevitable march of the combine harvester in all the wheat-producing areas of the world is one of the latest examples of the mechanization of the agriculture of the West. Cultivation tends to be quicker and deeper. There is a growing feeling that the more and the deeper the soil is stirred the better will be the crop. The invention of the gyrotiller, a heavy and expensive soil churn, is one of the answers to this demand. The slaves of the Roman Empire have been replaced by mechanical slaves. The replacement of the horse and the ox by the internal combustion engine and the electric motor is, however, attended by one great disadvantage. These machines do not void urine and dung and so contribute nothing to the maintenance of soil fertility. In this sense the slaves of Western agriculture are less efficient than those of ancient Rome.

Artificial manures are widely used. The feature of the manuring of the West is the use of artificial manures. The factories engaged during the Great War in the fixation of atmospheric nitrogen for the manufacture of explosives had to find other markets, the use of nitrogenous fertilizers in agriculture increased, until to-day the majority of farmers and market gardeners base their manurial programme on the cheapest forms of nitrogen (N), phosphorus (P), and potassium (K) on the market. What may be conveniently described as the NPK mentality dominates farming alike in the experimental stations and the country-side. Vested interests, entrenched in time of national emergency, have gained a stranglehold.

Artificial manures involve less labour and less trouble than farm-yard manure. The tractor is superior to the horse in power and in speed of work: it needs no food and no expensive care during its long hours of rest. These two agencies have made it easier to run a farm. A satisfactory profit and loss account has been obtained. For the moment farming has been made to pay. But there is another side to this picture. These chemicals and these machines can do nothing to keep the soil in good heart. By their use the processes of growth can never be balanced by the processes of decay. All that they can accomplish is the transfer of the soil's capital to current account. That this is so will be much clearer when the attempts now being made to farm without any animals at all march to their inevitable failure.

Diseases are on the increase. With the spread of artificials and the exhaustion of the original supplies of humus, carried by every fertile soil, there has been a corresponding increase in the diseases of crops and of the animals which feed on them. If the spread of foot-and-mouth disease in Europe and its comparative insignificance among well fed animals in the East are compared, or if the comparison is made between certain areas in Europe, the conclusion is inevitable that there must be an intimate connexion between faulty methods of agriculture and animal disease. In crops like potatoes and fruit, the use of the poison spray has closely followed the reduction in the supplies of farm-yard manure and the diminution of fertility.

Food preservation processes are also on the increase. A feature of the agriculture of the West is the development of food preservation processes by which the journey of products like meat, milk, vegetables, and fruit between the soil and the stomach is prolonged. This is done by freezing, by the use of carbon dioxide, by drying, and by canning. Although food is preserved for a time in this way, what is the effect of these processes on the health of the community during a period of, say, twenty-five years? Is it possible to preserve the first freshness of food? If so then science will have made a very real contribution.

Science has been called in to help production. Another of the features of the agriculture of the West is the development of agricultural science. Efforts have been made to enlist the help of a number of separate sciences in studying the problems of agriculture and in increasing the production of the soil. This has entailed the foundation of numerous experiment stations which every year pour out a large volume of advice in the shape of printed matter.

These mushroom ideas of agriculture are failing; mother earth deprived of her manurial rights is in revolt; the land is going on strike; the fertility of the soil is declining. An examination of the areas which feed the population and the machines of a country like Great Britain leaves no doubt that the soil is no longer able to stand the strain. Soil fertility is rapidly diminishing, particularly in the United States, Canada, Africa, Australia, and New Zealand. In Great Britain itself real farming has already been given up except on the best lands. The loss of fertility all over the world is indicated by the growing menace of soil erosion. The seriousness of the situation is proved by the attention now being paid to this matter in the press and by the various Administrations. In the United States, for example, the whole resources of government are being mobilized to save what is left of the good earth.

The agricultural record has been briefly reviewed from the standpoint of soil fertility. The main characteristics of the various methods of agriculture have been summarized. The most significant of these are the operations of Nature as seen in the forest. There the fullest use is made of sunlight and rainfall in raising heavy crops of produce and at the same time not only maintaining fertility but actually building up large reserves of humus. The peasants of China, who pay great attention to the return of all wastes to the land, come nearest to the ideal set by Nature. They have maintained a large population on the land without any falling off in fertility. The agriculture of ancient Rome failed because it was unable to maintain the soil in a fertile condition. The farmers of the West are repeating the mistakes made by Imperial Rome. The soils of the Roman Empire, however, were only called upon to assuage the hunger of a relatively small population. The demands of the machine were then almost non-existent. In the West there are relatively more stomachs to fill while the growing hunger of the machine is an additional burden on the soil. The Roman Empire lasted for eleven centuries. How long will the supremacy of the West endure? The answer depends on the wisdom and courage of the population in dealing with the things that matter. Can mankind regulate its affairs so that its chief possession—the fertility of the soil—is preserved? On the answer to this question the future of civilization depends.

BIBLIOGRAPHY

Agricultural Statistics of India, I, Delhi, 1938.

HOWARD, A., and HOWARD, G. L. C. *The Development of Indian Agriculture*, Oxford University Press, 1929.

KING, F. H. *Farmers of Forty Centuries or Permanent Agriculture in China, Korea, and Japan*, London, 1926.

LYMINGTON, VISCOUNT. *Famine in England*, London, 1938.

MOMMSEN, THEODOR. *The History of Rome*, transl. Dickson, London, 1894.

WRENCH, G. T. *The Wheel of Health*, London, 1938.

PART I

THE PART PLAYED BY SOIL FERTILITY IN AGRICULTURE

CHAPTER II

THE NATURE OF SOIL FERTILITY

WHAT is this soil fertility? What exactly does it mean? How does it affect the soil, the crop, and the animal? How can we best investigate it? An attempt will be made in this chapter to answer these questions and to show why soil fertility must be the basis of any permanent system of agriculture.

The nature of soil fertility can only be understood if it is considered in relation to Nature's round. In this study we must at the outset emancipate ourselves from the conventional approach to agricultural problems by means of the separate sciences and above all from the statistical consideration of the evidence afforded by the ordinary field experiment. Instead of breaking up the subject into fragments and studying agriculture in piecemeal fashion by the analytical methods of science, appropriate only to the discovery of new facts, we must adopt a synthetic approach and look at the wheel of life as one great subject and not as if it were a patchwork of unrelated things.

All the phases of the life cycle are closely connected; all are integral to Nature's activity; all are equally important; none can be omitted. We have therefore to study soil fertility in relation to a natural working system and to adopt methods of investigation in strict relation to such a subject. We need not strive after quantitative results: the qualitative will often serve. We must look at soil fertility as we would study a business where the profit and loss account must be taken along with the balance-sheet, the standing of the concern, and the method of management. It is the 'altogetherness' which matters in business, not some particular transaction or the profit or loss of the current year. So it is with soil fertility. We have to consider the wood, not the individual trees.

The wheel of life is made up of two processes—growth and decay. The one is the counterpart of the other.

Let us first consider growth. The soil yields crops; these form the food of animals: crops and animals are taken up into the human body and are digested there. The perfectly grown, normal, vigorous human being is the highest natural development known to us. There is no break in the chain from soil to man; this section of the wheel of life is uninterrupted throughout; it is also an integration; each step depends on the last. It must therefore be studied as a working whole.

The energy for the machinery of growth is derived from the sun; the chlorophyll in the green leaf is the mechanism by which this energy is intercepted; the plant is thereby enabled to manufacture food—to synthesize carbohydrates and proteins from the water and other substances taken up by the roots and the carbon dioxide of the atmosphere. The efficiency of the green leaf is therefore of supreme importance; on it depends the food supply of this planet, our well-being, and our activities. There is no alternative source of nutriment. Without sunlight and the green leaf our industries, our trade, and our possessions would soon be useless.

The chief factors on which the work of the green leaf depends are the condition of the soil and its relation to the roots of the plant. The plant and the soil come into gear by means of the root system in two ways—by the root hairs and by the mycorrhizal association. The first condition for this gearing is that the internal surface of the soil—the pore space—shall be as large as possible throughout the life of the crop. It is on the walls of this pore space, which are covered with thin water films, that the essential activities of the soil take place. The soil population, consisting mainly of bacteria, fungi, and protozoa, carry on their life histories in these water films.

The contact between the soil and the plant which is best understood takes place by means of the root hairs. These are prolongations of the outer layer of cells of the young root. Their duty is to absorb from the thin films of moisture on the walls of the pore space the water and dissolved salts needed for the work of the green leaves: no actual food can reach the plant in this way, only simple things which are needed by the green leaf to synthesize food. The activities of the pore space depend on respiration for which adequate quantities of oxygen are essential. A corresponding amount of carbon dioxide is the natural by-product. To maintain the oxygen supply and to reduce the amount of carbon dioxide, the pore spaces must be kept in contact with the atmosphere. The soil must be ventilated. Hence the importance of cultivation.

As most of the soil organisms possess no chlorophyll, and, moreover, have to work in the dark, they must be supplied with energy. This is obtained by the oxidation of humus—the name given to a complex residue of partly oxidized vegetable and animal matter together with the substances synthesized by the fungi and bacteria which break down these wastes. This humus also helps to provide the cement which enables the minute mineral soil particles to aggregate into larger compound particles and so maintain the pore space. If the soil is deficient in humus, the volume of the pore space is reduced; the aeration of the soil is impeded; there is insufficient organic matter for the soil population; the machinery of the soil runs down; the supply of oxygen, water, and dissolved salts needed by the root hairs is reduced; the synthesis of carbohydrates and proteins in the green leaf proceeds at a lower tempo; growth is affected. Humus is therefore an essential material for the soil if the first phase of the life cycle is to function.

There is another reason why humus is important. Its presence in the soil is an essential condition for the proper functioning of the second contact between soil and plant—the mycorrhizal relationship. By means of this connexion certain soil fungi, which live on humus, are able to invade the living cells of the young roots and establish an intimate relation with the plant, the details of which symbiosis are still being investigated and discussed. Soil fungus and plant cells live together in closer partnership than the algal and fungous constituents of the lichen do. How the fungus benefits has yet to be determined. How the plant profits is easier to understand. If a suitable preparation of such roots is examined under the microscope, all stages in the digestion of the fungous mycelium can be seen. At the end of the partnership the root consumes the fungus and in this manner is able to absorb the carbohydrates and proteins which the fungus obtains partly from the humus in the soil. The mycorrhizal association therefore is the living bridge by which a fertile soil (one rich in humus) and the crop are directly connected and by which food materials ready for immediate use can be transferred from soil to plant. How this association influences the work of the green leaf is one of the most interesting problems science has now to investigate. Is the effective synthesis of carbohydrates and proteins in the green leaf dependent on the digestion products of these soil fungi? It is more than probable that this must prove to be the case. Are these digestion products at the root of disease resistance and quality? It would appear so. If this is the case it would follow that on the efficiency of this mycorrhizal association the health and well-being of mankind must depend.

In a fertile soil the soil and the plant come into gear in two ways simultaneously. In establishing and maintaining these contacts humus is essential. It is therefore a key material in the life cycle. Without this substance the wheel of life cannot function effectively.

The processes of decay which round off and complete the wheel of life can be seen in operation on the floor of any woodland. This has already been discussed (p. 2). It has been shown how the mixed animal and vegetable wastes are converted into humus and how the forest manures itself.

Such are the essential facts in the wheel of life. Growth on the one side: decay on the other. In Nature's farming a balance is struck and maintained between these two complementary processes. The only man-made systems of agriculture—those to be found in the East—which have stood the test of time have faithfully copied this rule in Nature. It follows therefore that the correct relation between the processes of growth and the processes of decay is the first principle of successful farming. Agriculture must always be balanced. If we speed up growth we must accelerate decay. If, on the other hand, the soil's reserves are squandered, crop production ceases to be good farming: it becomes something very different. The farmer is transformed into a bandit.

It is now possible to define more clearly the meaning of soil fertility. It is the condition of a soil rich in humus in which the growth processes proceed rapidly, smoothly, and efficiently. The term therefore connotes such things as abundance, high quality, and resistance to disease. A soil which grows to perfection a wheat crop—the food of man—is described as fertile. A pasture on which meat and milk of the first class are produced falls into the same category. An area under market-garden crops on which vegetables of the highest quality are raised has reached the peak as regards fertility.

Why does soil fertility so markedly influence the soil, the plant, and the animal? By virtue of the humus it contains. The nature and properties of this substance as well as the products of its decomposition are therefore important. These matters must now be considered.

What is humus? A reply to this question has been rendered easier by the appearance in 1938 of the second edition of Waksman's admirable monograph on humus in which the results of no less than 1311 original papers have been reduced to order. Waksman defines humus as

'a complex aggregate of brown to dark-coloured amorphous substances, which have originated during the decomposition of plant and animal residues by micro-organisms, under aerobic and anaerobic conditions, usually in soils, composts, peat bogs, and water basins. Chemically, humus consists of various constituents of the original plant material resistant to further decomposition; of substances undergoing decomposition; of complexes resulting from decomposition either by processes of hydrolysis or by oxidation and reduction; and of various compounds synthesized by micro-organisms. Humus is a natural body; it is a composite entity, just as are plant, animal, and microbial substances; it is even much more complex chemically, since all these materials contribute to its formation. Humus possesses certain specific physical, chemical, and biological properties which make it distinct from other natural organic bodies. Humus, in itself or by interaction with certain inorganic constituents of the soil, forms a complex colloidal system, the different constituents of which are held together by surface forces; this system is adaptable to changing conditions of reaction, moisture, and action by electrolytes. The numerous activities of the soil micro-organisms take place in this system to a large extent.'

Viewed from the standpoint of chemistry and physics humus is therefore not a simple substance: it is made up from a group of very complex organic compounds depending on the nature of the residues from which it is formed, on the conditions under which decomposition takes place, and on the extent to which the processes of decay have proceeded. Humus, therefore, cannot be exactly the same thing everywhere. It is bound to be a creature of circumstance. Moreover it is alive and teems with a vast range of micro-organisms which derive most of their nutriment from this substratum. Humus in the natural state is dynamic, not static. From the point of view of agriculture, therefore, we are dealing not with simple dead matter like a sack of sulphate of ammonia, which can be analysed and valued according to its chemical composition, but with a vast organic complex in which an important section of the farmer's invisible labour force—the organisms which carry on the work of the soil—is temporarily housed. Humus, therefore, involves the element of labour; in this respect also it is one of the most important factors on the farm.

It is essential at this point to pay some attention to the many-sided properties of humus and to realize how profoundly it differs from a chemical manure. At the moment all over the world field trials—based on mere nitrogen content—are in progress for comparing, on the current crop, dressings of humus and various artificial manures. A mere glance at the properties of humus will show that such field trials are based on a fundamental misconception of what soil fertility implies and are misleading and therefore useless.

The properties of humus have been summed up by Waksman as follows:

'1. Humus possesses a dark brown to black colour.

'2. Humus is practically insoluble in water, although a part of it may go into colloidal solution in pure water. Humus dissolves to a large extent in dilute alkali solutions, especially on boiling, giving a dark coloured extract; a large part of this extract precipitates when the alkali solution is neutralized by mineral acids.

'3. Humus contains a somewhat larger amount of carbon than do plant, animal, and microbial bodies; the carbon content of humus is usually about 55 to 56 per cent., and frequently reaches 58 per cent.

'4. Humus contains considerable nitrogen, usually about 3 to 6 per cent. The nitrogen concentration may be frequently less than this figure; in the case of certain high-moor peats, for example, it may be only 0·5–0·8 per cent. It may also be higher, especially in sub-soils, frequently reaching 10 to 12 per cent.

'5. Humus contains the elements carbon and nitrogen in proportions which are close to 10:1; this is true of many soils and of humus in sea bottoms. This ratio varies considerably with the nature of the humus, the stage of its decomposition, the nature and depth of soil from which it is obtained, the climatic and other environmental conditions under which it is formed.

'6. Humus is not in a static, but rather in a dynamic, condition, since it is constantly formed from plant and animal residues and is continuously decomposed further by micro-organisms.

'7. Humus serves as a source of energy for the development of various groups of micro-organisms, and during decomposition gives off a continuous stream of carbon dioxide and ammonia.

'8. Humus is characterized by a high capacity of base-exchange, of combining with various other soil constituents, of absorbing water, and of swelling, and by other physical and physico-chemical properties which make it a highly valuable constituent of substrates which support plant and animal life.'

To this list of properties must be added the role of humus as a cement in creating and maintaining the compound soil particles so important in the maintenance of tilth.

The effect of humus on the crop is nothing short of profound. The farmers and peasants who live in close touch with Nature can tell by a glance at the crop whether or not the soil is rich in humus. The habit of the plant then develops something approaching personality; the foliage assumes a characteristic set; the leaves acquire the glow of health; the flowers develop depth of colour; the minute morphological characters of the whole of the plant organs become clearer and sharper. Root development is profuse: the active roots exhibit not only turgidity but bloom.

The influence of humus on the plant is not confined to the outward appearance of the various organs. The quality of the produce is also affected. Seeds are better developed, and so yield better crops and also provide live stock with a satisfaction not conferred by the produce of worn-out land. The animals need less food if it comes from fertile soil. Vegetables and fruit grown on land rich in humus are always superior in quality, taste, and keeping power to those raised by other means. The quality of wines, other things being equal, follows the same rule. Almost every villager in countries like France appreciates these points and will talk of them freely without the slightest prompting.

In the case of fodder a very interesting example of the relation between soil fertility and quality has recently been investigated. This was noticed in the meadows of La Crau between Salon and Arles in Provence. Here the fields are irrigated with muddy water, containing finely divided limestone drawn from the Durance, and manured mostly with farm-yard manure. The soils are open and permeable, the land is well drained naturally. All the factors on which soil fertility depends are present together—an open soil with ample organic matter, ample moisture, and the ideal climate for growth. Any grazier who saw these meadows for the first time would at once be impressed by them: a walk through the fields at hay-making would prepare him for the news that it pays the owners of high-quality animals to obtain their roughage from this distant source. Several cuts of hay are produced every year, which enjoy such a reputation for quality that the bales are sent long distances by motor lorry to the various racing stables of France and are even exported to Newmarket. The small stomach of the racehorse needs the very best food possible. This the meadows of La Crau help to produce.

The origin of these irrigated meadows would provide an interesting story. Did they arise as the result of a set of permanent manurial experiments on the Broadbalk model or through the work of some observant local pioneer? I suspect the second alternative will be found to be nearer the truth. A definite answer to this question is desirable because in a recent discussion at Rothamsted, on the relation between a fertile soil and high-quality produce, it was stated that no evidence of such a connexion could be discovered in the literature. The farmers of Provence, however, have supplied it and also a measure of quality in the shape of a satisfactory price. For the present the only way of measuring quality seems to be by selling it. It cannot be weighed and measured by the methods of the laboratory. Nevertheless it exists: moreover it constitutes a very important factor in agriculture. Apparently some of the experiment stations have not yet come to grips with this factor: the farmers have. The sooner therefore that effective liaison is established between these two agencies the better.

The effect of soil fertility on live stock can be observed in the field. As animals live on crops we should naturally expect the character of the plant as regards nutrition to be passed on to stock. This is so. The effect of a fertile soil can at once be seen in the condition of the animals. This is perhaps most easily observed in the bullocks fattened on some of the notable pastures in Great Britain. The animals show a well-developed bloom, the coat and skin look and feel right, the eyes are clear, bright, and lively. The posture of the animal betokens health and well-being. It is not necessary to weigh or measure them. A glance on the part of a successful grazier, or of a butcher accustomed to deal with high-class animals, is sufficient to tell them whether all is well or whether there is something wrong with the soil or the management of the animals or both. The results of a fertile soil and proper methods of management are measured by the prices these animals fetch in the market and the standing of the farmer in these markets. It should be a compulsory item in the training of agricultural investigators to accompany some of the best of our English cattle from the pasture to the market and watch what happens there. They would at once discover that the most fertile pastures produce the best animals, that auctioneers and buyers detect quality instantly, and that such animals find a ready sale and command the best prices. The reputation of the pastures is finally passed on to the butcher and to his clients.

Resistance to insect and fungous disease is also conferred by humus. Perhaps the best examples of this are to be seen in the East. In India, the crops grown on the highly fertile soils round the 500,000 villages suffer remarkably little from pests. This subject is developed at length in a future chapter (p. 156) when the retreat of the crop and of the animal before the parasite is discussed.

Soil fertility not only influences crops and live stock but also the fauna of the locality. This is perhaps most easily seen in the fish of streams which flow through areas of widely differing degrees of fertility. An example of such difference is referred to at the end of Chapter V of Isaac Walton's *Compleat Angler* in the following words:

'And so I shall proceed next to tell you, it is certain, that certain fields near Leominster, a town in Herefordshire, are observed to make sheep that graze upon them more fat than the next, and also to bear finer wool; that is to say, that in that year in which they feed in such a particular pasture, they shall yield finer wool than they did that year before they came to feed in it, and coarser again if they shall return to their former pasture; and again return to a finer wool, being fed in the fine wool ground. Which I tell you, that you may the better believe that I am certain, if I catch a trout in one meadow he shall be white and faint, and very likely to be lousy; and as certainly if I catch a trout in the next meadow, he shall be strong and red and lusty and much better meat: trust me, scholar. I have caught many a trout in a particular meadow, that the very shape and enamelled colour of him hath been such as hath joyed me to look on him: and I have then with much pleasure concluded with Solomon, "Everything is beautiful in his season".'

Soil fertility is the condition which results from the operation of Nature's round, from the orderly revolution of the wheel of life, from the adoption and faithful execution of the first principle of agriculture—there must always be a perfect balance between the processes of growth and the processes of decay. The consequences of this condition are a living soil, abundant crops of good quality, and live stock which possess the bloom of health. The key to a fertile soil and a prosperous agriculture is humus.

BIBLIOGRAPHY

RAYNER, M. C. *Mycorrhiza: an Account of Non-pathogenic Infection by Fungi in Vascular Plants and Bryophytes*, London, 1927.
—— 'Mycorrhiza in relation to Forestry', *Forestry*, viii, 1934, p. 96; x, 1936, p. 1; and xiii, 1939, p. 19.
WAKSMAN, S. A. *Humus: Origin, Chemical Composition, and Importance in Nature*, London, 1938.

CHAPTER III

THE RESTORATION OF FERTILITY

T HE moment mankind undertook the business of raising crops and breeding animals, the processes of Nature were subjected to interference. Soil fertility was exploited for the growing of food and the production of the raw materials—such as wool, skins, and vegetable fibres—needed for clothing. Up to the dawn of the Industrial Revolution in the West, the losses of humus involved in these agricultural operations were made up either by the return of waste material to the soil or by taking up virgin land.

Where the return of wastes balanced the losses of humus involved in production, systems of agriculture became stabilized and there was no loss of fertility. The example of China has already been quoted. The old mixed farming of a large part of Europe, including Great Britain—characterized by a correct balance between arable and live stock, the conversion of wastes into farm-yard manure, methods of sheep folding, and the copious use of the temporary ley—is another instance of the same thing.

The constant exploitation of new areas to replace worn-out land has also gone on for centuries and is still taking place. Sometimes this has involved wars and conquests: at other times nothing more than taking up fresh prairie or forest land wherever this was to be found. A special method is adopted by some primitive tribes. The forest growth is burnt down, the store of humus is converted into crops, the exhausted land is given back to Nature for reafforestation and the building up of a new reserve of humus. In a rough and ready way fertility is maintained. Such shifting cultivation still exists all over the world, but like the taking up of new land is only possible when the population is small and suitable land abundant. This burning process has even been incorporated into permanent systems of agriculture and has proved of great value in rice cultivation in western India. Here the intractable soils of the rice nurseries have to be prepared during the last part of the hot season so that the seedlings are ready for transplanting by the break of the monsoon. This is achieved by covering the nurseries with branches collected from the forest and setting fire to the mass. The heat destroys the colloids, restores the tilth, and makes the manuring and cultivation of the rice nurseries possible.

It is an easy matter to destroy a balanced agriculture. Once the demand for food and raw materials increases and good prices are obtained for the produce of the soil, the pressure on soil fertility becomes intense. The temptation to convert this fertility into money becomes irresistible. Western agriculture was subjected to this strain by the very rapid developments which followed the invention of the steam-engine, the internal combustion engine, electrically driven motors, and improvements in communications and transport. Factory after factory arose; a demand for labour followed; the urban population increased. All these developments provided new and expanding markets for food and raw materials. These were supplied in three ways—by cashing-in the existing fertility of the whole world, by the use of a temporary substitute for soil fertility in the shape of artificial manures, and by a combination of both methods. The net result has been that agriculture has become unbalanced and therefore unstable.

Let us review briefly the operations of Western agriculture from the point of view of the utilization of wastes in order to discover whether the gap between the losses and gains of humus, now bridged by artificials, can be reduced or abolished altogether. If this is possible, something can be done to restore the balance of agriculture and to make it more stable and therefore more permanent.

Many sources of soil organic matter exist, namely: (1) the roots of crops, weeds, and crop residues which are turned under in the course of cultivation; (2) the algae met with in the surface soil; (3) temporary leys, the turf of worn-out grass land, catch crops, and green-manures; (4) the urine of animals; (5) farmyard manure; (6) the contents of the dustbins of our cities and towns; (7) certain factory wastes which result from the processing of agricultural produce; (8) the wastes of the urban population; (9) water-weeds, including seaweed. These must now be very briefly considered. In later chapters most of these matters will be referred to again and discussed in greater detail.

The residues turned under in the course of cultivation. It is not always realized that about half of every crop—the root system—remains

D

in the ground at harvest time and thus provides a continuous return of organic matter to the soil. The weeds and their roots ploughed in during the ordinary course of cultivation add to this supply. When these residues, supplemented by the fixation of nitrogen from the atmosphere, are accompanied by skilful soil management, which safeguards the precious store of humus, crop production can be maintained at a low level without the addition of any manure whatsoever beyond the occasional droppings of live stock and birds. A good example of such a system of farming without manure is to be found in the alluvial soils of the United Provinces in India where the field records of ten centuries prove that the land produces small crops year after year without any falling off in fertility. A perfect balance has been reached between the manurial requirements of the crops harvested and the natural processes which recuperate fertility. The greatest care, however, is taken not to over-cultivate, not to cultivate at the wrong time, or to stimulate the soil processes by chemical manures. Systems of farming such as these supply as it were the base-line for agricultural development. A similar though not so convincing result is provided by the permanent wheat plot at Rothamsted, where this crop has been grown on the same land without manure since 1844. This plot, which has been without manure of any kind since 1839, showed a slow decline in production for the first eighteen years, after which the yield has been practically constant. The reserves of humus in this case left over from the days of mixed farming evidently lasted for nearly twenty years. There are, however, two obvious weaknesses in this experiment. This plot does not represent any system of agriculture, it only speaks for itself. Nothing has been done to prevent earthworms and other animals from bringing in a constant supply of manure, in the shape of their wastes, from the surrounding land. It is much too small to yield a significant result.

Soil algae are a much more important factor in the tropics than in temperate regions. Nevertheless they occur in all soils and often play a part in the maintenance of soil fertility. Towards the end of the rainy season in countries like India a thick algal film occurs on the surface of the soil which immobilizes a large amount of combined nitrogen otherwise likely to be lost by leaching.

While this film is forming cultivation is suspended and weeds are allowed to grow. Just before the sowing of the cold weather crops in October the land is thoroughly cultivated, when this easily decomposable and finely divided organic matter, which is rich in nitrogen, is transformed into humus and then into nitrates. How far a similar method can be utilized in colder countries is a matter for investigation. In the East cultivation always fits in with the life-cycle in a remarkable way. In the West cultivation is regarded as an end in itself and not, as it should be, as a factor in the wheel of life. Europe has much to learn from Asia in the cultivation of the soil.

Temporary leys, catch-crops, green-manures, and the turf of worn-out grass land are perhaps the most important source of humus in Western agriculture. All these crops develop a large root system; the permanent and temporary leys give rise to ample residues of organic matter which accumulate in the surface soil. Green-manures and catch-crops develop a certain amount of soft and easily decomposable tissue. Provided these crops are properly utilized a large addition of new humus can be added to the soil. The efficiency of these methods of maintaining soil fertility could, however, be very greatly increased.

The urine of animals. The key substance in the manufacture of humus from vegetable wastes is urine—the drainage of the active cells and glands of the animal. It contains in a soluble and balanced form all the nitrogen and minerals, and in all probability the accessory growth-substances as well, needed for the work of the fungi and bacteria which break down the various forms of cellulose—the first step in the synthesis of humus. It carries in all probability every raw material, known and unknown, discovered and undiscovered, needed in the building up of a fertile soil. Much of this vital substance for restoring soil fertility is either wasted or only imperfectly utilized. This fact alone would explain the disintegration of the agriculture of the West.

Although *farm-yard manure* has always been one of the principal means of replenishing soil losses, even now the methods by which this substance is prepared are nothing short of deplorable. The making of farm-yard manure is the weakest link in the agriculture of Western countries. For centuries this weakness has been the fundamental fault of Western farming, one completely overlooked by many observers and the great majority of investigators.

Dustbin refuse. Practically no agricultural use is now being made of the impure cellulose and kitchen wastes which find their way into the urban dustbin. These are mostly buried in controlled tips or burnt.

Animal residues. A number of wastes connected with the processing of food and some of the raw materials needed in industry are utilized on the land and find a ready market. The animal residues include such materials as dried blood, feathers, greaves, hair waste, hoof and horn, rabbit waste, slaughter-house refuse, and fish waste. There is a brisk demand for most of these substances, as they give good results on the land. The only drawback is the limited supplies available. The organic residues from manufacture consist of damaged oil-cakes, shoddy and tannery waste, of which shoddy, a by-product of the wool industry, is the most important. These two classes of wastes, animal and industrial, are applied to the soil direct and, generally speaking, command much higher prices than would be expected from their content of nitrogen, phosphorus, and potash. This is because the soil is in such urgent need of humus and because the supply falls so far short of the demand. It is probable that a better use for these wastes will be found as raw materials for the compost heaps of the future, where they will act as substitutes for urine in the breaking down of dustbin refuse in localities where the supply of farm-yard manure is restricted.

Water weeds. Little use is made of water weeds in maintaining soil fertility. Perhaps the most useful of these is seaweed, which is thrown up on the beaches in large quantities at certain times of the year and which contains iodine and includes the animal residues needed for converting vegetable wastes into humus. Many of our sea-side resorts could easily manufacture from seaweed and dustbin refuse the vast quantities of humus needed for the farms and market gardens in their neighbourhood and so balance the local agriculture. Little or nothing, however, is being done in this direction. In some cases the seaweed collection on pleasure beaches is taken up by the farmers with good results, but the systematic utilization of seaweed in the compost heap is still a matter for the future. The streams and rivers which carry off the surplus rainfall also contain appreciable quantities of combined nitrogen and minerals in solution. Much of this could be intercepted by the cultivation of suitable plants on the borders of these streams which would furnish large quantities of easily decomposable material for humus manufacture.

The *night soil and urine* of the population is at present almost completely lost to the land. In urban areas the concentration of the population is the main reason why water-borne sewage systems have developed. The greatest difficulty in the path of the reformer is the absence of sufficient land for dealing with these wastes. In country districts, however, there are no insurmountable obstacles to the utilization of human wastes.

It will be evident that in almost every case the vegetable and animal residues of Western agriculture are either being completely wasted or else imperfectly utilized. A wide gap between the humus used up in crop production and the humus added as manure has naturally developed. This has been filled by chemical manures. The principle followed, based on the Liebig tradition, is that any deficiencies in the soil solution can be made up by the addition of suitable chemicals. This is based on a complete misconception of plant nutrition. It is superficial and fundamentally unsound. It takes no account of the life of the soil, including the mycorrhizal association—the living fungous bridge which connects soil and sap. Artificial manures lead inevitably to artificial nutrition, artificial food, artificial animals, and finally to artificial men and women.

The ease with which crops can be grown with chemicals has made the correct utilization of wastes much more difficult. If a cheap substitute for humus exists why not use it? The answer is twofold. In the first place, chemicals can never be a substitute for humus because Nature has ordained that the soil must live and the mycorrhizal association must be an essential link in plant nutrition. In the second place, the use of such a substitute cannot be cheap because soil fertility—one of the most important assets of any country—is lost; because artificial plants, artificial animals, and artificial men are unhealthy and can only be protected from the parasites, whose duty it is to remove them, by means of poison sprays, vaccines and serums and an expensive system of patent medicines, panel doctors, hospitals, and so forth. When the finance of crop production is considered together with that of the various social services which are needed to repair the consequences of an unsound agriculture, and when it is borne in mind that our greatest possession is a healthy, virile population, the cheapness of artificial manures disappears altogether. In the years to come chemical manures will be considered as one of the greatest follies of the industrial epoch. The teachings of the agricultural economists of this period will be dismissed as superficial.

In the next section of this book the methods by which the agriculture of the West can be reformed and balanced and the use of artificial manures given up will be discussed.

BIBLIOGRAPHY

Clarke, G. 'Some Aspects of Soil Improvement in relation to Crop Production', *Proc. of the Seventeenth Indian Science Congress*, Asiatic Society of Bengal, Calcutta, 1930, p. 23.

Hall, Sir A. Daniel. *The Improvement of Native Agriculture in relation to Population and Public Health*, Oxford University Press, 1936.

Howard, A., and Wad, Y. D. *The Waste Products of Agriculture: their Utilization as Humus*, Oxford University Press, 1931.

Mann, H. H., Joshi, N. V., and Kanitkar, N. V. 'The *Rab* System of Rice Cultivation in Western India', *Mem. of the Dept. of Agriculture in India (Chemical Series)*, ii, 1912, p. 141.

Manures and Manuring, Bulletin 36 of the Ministry of Agriculture and Fisheries, H.M. Stationery Office, 1937.

PART II

THE INDORE PROCESS

CHAPTER IV
THE INDORE PROCESS

THE Indore Process for the manufacture of humus from vegetable and animal wastes was devised at the Institute of Plant Industry, Indore, Central India, between the years 1924 and 1931. It was named after the Indian State in which it originated, in grateful remembrance of all the Indore Darbar did to make my task in Central India easier and more pleasant.

Although the working out of the actual process only took seven years, the foundations on which it is based occupied me for more than a quarter of a century. Two independent lines of thought and study led up to the final result. One of these concerns the nature of disease and is discussed more fully in Chapter XI under the heading—'The Retreat of the Crop and the Animal before the Parasite'. It was observed in the course of these studies that the maintenance of soil fertility is the real basis of health and of resistance to disease. The various parasites were found to be only secondary matters: their activities resulted from the breakdown of a complex biological system—the soil in its relation to the plant and to the animal—due to improper methods of agriculture, an impoverished soil, or to a combination of both.

The second line of thought arose in the course of nineteen years (1905–24) spent in plant-breeding at Pusa, when it was gradually realized that the full possibilities of the improvement of the variety can only be achieved when the soil in which the new types are grown is provided with an adequate supply of humus. Improved varieties by themselves could be relied upon to give an increased yield in the neighbourhood of 10 per cent.: improved varieties plus better soil conditions were found to produce an increment up to 100 per cent. or even more. As an addition of even 10 per cent. to the yield would ultimately impose a severe strain on the frail fertility reserves of the soils of India and would gradually lead to their impoverishment, plant-breeding to achieve any permanent

success would have to include a continuous addition to the
humus content of the small fields of the Indian cultivators.
The real problem was not the improvement of the variety but
how simultaneously to make the variety and the soil more
efficient.

By about the year 1918 these two hitherto independent ap-
proaches to the problems of crop production—by way of pathology
and by way of plant-breeding—began to coalesce. It became
clearer and clearer that agricultural research itself was involved
in the problem; that the organization was responsible for the
failure to recognize the things that matter in agriculture and would
therefore have to be reformed; the separation of work on crops
into such compartments as plant-breeding, mycology, entomology,
and so forth, would have to be given up; the plant would have to
be studied in relation to the soil on the one hand and to the agricultural practices of the locality on the other. An approach to the
problems of crop production on such a wide front was obviously
impossible in a research institute like Pusa in which the work on
crops was divided into no less than six separate sections. The
working out of a method of manufacturing humus from waste
products and a study of the reaction of the crop to improved soil
conditions would encroach on the work of practically every section
of the Institute. As no progress has ever been made in science
without complete freedom, the only way of studying soil fertility
as one subject appeared to be to found a new institute in which the
plant would be the centre of the subject and where science and
practice could be brought to bear on the problem without any
consideration of the existing organization of agricultural research.
Thanks to the support of a group of Central Indian States and a
large grant from the Indian Central Cotton Committee, the Institute of Plant Industry was founded at Indore in 1924. Central
India was selected as the home of this new research centre for two
reasons: (1) the offer on a 99 years' lease of an area of 300 acres of
suitable land by the Indore Darbar, and (2) the absence in the
Central India Agency of any organized system of agricultural
research such as had been established throughout British India.
This tract therefore provided the land on the one hand and
freedom from interference on the other for the working out of a

new approach, based on the humus content of the soil, to the problems underlying crop production.[1]

The work at Indore accomplished two things: (1) the obsolete character of the present-day organization of agricultural research was demonstrated; (2) a practical method of manufacturing humus was devised.

The Indore Process was first described in detail in 1931 in Chapter IV of *The Waste Products of Agriculture*. Since that date the method has been taken up by most of the plantation industries and also on many farms and gardens all over the world. In the course of this work nothing has been added to the two main principles underlying the process, namely, (1) the admixture of vegetable and animal wastes with a base for neutralizing acidity, and (2) the management of the mass so that the micro-organisms which do the work can function in the most effective manner. A number of minor changes in working have, however, been suggested. Some of these have proved advantageous in increasing the output. In the following account the original description has been followed, but all useful improvements have been incorporated: the technique has been brought up to date.

THE RAW MATERIALS NEEDED

1. *Vegetable Wastes.* In temperate countries like Great Britain these include—straw, chaff, damaged hay and clover, hedge and bank trimmings, weeds including sea- and water-weeds, prunings, hop-bine and hop-string, potato haulm, market-garden residues including those of the greenhouse, bracken, fallen leaves, sawdust, and wood shavings. A limited amount of other vegetable material like the husks of cotton seed, cacao, and ground nuts as well as banana stalks are also available near some of the large cities.

In the tropics and sub-tropics the vegetable wastes consist of very similar materials including the vegetation of waste areas, grass, plants grown for shade and green-manure, sugar-cane leaves and stumps, all crop residues not consumed by live stock, cotton stalks, weeds, sawdust and wood shavings, and plants grown for providing compostable material on the borders of fields, roadsides, and any vacant corners available.

[1] An account of the organization of the Institute of Plant Industry was published as *The Application of Science to Crop-Production* by the Oxford University Press in 1929.

A continuous supply of mixed dry vegetable wastes throughout the year, in a proper state of division, is the chief factor in the process. The ideal chemical composition of these materials should be such that, after being used as bedding for live stock, the carbon: nitrogen ratio is in the neighbourhood of 33:1. The material should also be in such a physical condition that the fungi and bacteria can obtain ready access to and break down the tissues without delay. The bark, which is the natural protection of the celluloses and lignins against the inroads of fungi, must first be destroyed. This is the reason why all woody materials—such as cotton and pigeon-pea stalks—were always laid on the roads at Indore and crushed by the traffic into a fine state of division before composting.

All over the world one of the first objections to the adoption of the Indore Process is that there is nothing worth composting or only small supplies of such material. In practically all such cases any shortage of wastes has soon been met by a more effective use of the land and by actually growing plants for composting on every possible square foot of soil. If Nature's way of using sunlight to the full in the virgin forest is compared with that on the average farm or on the average tea and rubber estate, it will be seen what leeway can be made up in growing suitable material for making humus. Sometimes the objection is heard that all this will cost too much. The answer is provided by the dust-bowls of North America. The soil must have its manurial rights or farming dies.

2. *Animal Residues.* The animal residues ordinarily available all over the world are much the same—the urine and dung of live stock, the droppings of poultry, kitchen waste including bones. Where no live stock is kept and animal residues are not available, substitutes such as dried blood, slaughter-house refuse, powdered hoof and horn, fish manure, and so forth can be employed. The waste products of the animal in some form or another are essential if real humus is to be made for the two following reasons.

(*a*) The verdict given by mother earth between humus made with animal residues and humus made with chemical activators like calcium cyanamide and the various salts of ammonia has always been in favour of the former. One has only to feel and smell a handful of compost made by these two methods to understand the plant's preference for humus made with animal residues. The one is soft to the feel with the smell of rich woodland earth: the other is often harsh to the touch with a sour odour. Sometimes when the two samples of humus made from similar vegetable wastes are analysed, the better report is obtained by the compost made with chemical activators. When, however, they are applied to the soil the plant speedily reverses the verdict of the laboratory. Dr. Rayner refers to this conflict between mother earth and the analyst, in the case of some composts suitable for forestry nurseries, in the following words:

'Full chemical analyses are now available for a number of these composts, and it is not without interest to recall that in the initial stages of the work a competent critic reported on one of them—since proved to be among the most effective—on a basis of comparative analysis, as "an organic manure of comparatively little value"; while another—since proved least successful of all those tested—was approved as a "first-class organic manure".'

The activator used in the first case was dried blood, in the second case an ammonium salt.

(b) No permanent or effective system of agriculture has ever been devised without the animal. Many attempts have been made, but sooner or later they break down. The replacement of live stock by artificials is always followed by disease the moment the original store of soil fertility is exhausted.

Where live stock is maintained the collection of their waste products—urine and dung—in the most effective manner is important.

At Indore the work-cattle were kept in well-ventilated sheds with earthen floors and were bedded down daily with mixed vegetable wastes including about 5 per cent. by volume of hard resistant material such as wood shavings and sawdust. The cattle slept on this bedding during the night when it was still further broken up and impregnated with urine. Next morning the soiled bedding and cattle dung were removed to the pits for composting; the earthen floor was then swept clean and all wet places were covered with new earth, after scraping out the very wet patches. In this way all the urine of the animals was absorbed; all smell in the cattle sheds was avoided, and the breeding of flies in the earth underneath the animals was entirely prevented. A new layer of bedding for the next day was then laid.

Every three months the earth under the cattle was changed, the urine-impregnated soil was broken up in a mortar mill and stored under cover near the compost pits. This urine earth, mixed with any wood ashes available, served as a combined activator and base in composting.

In the tropics, where there is abundance of labour, no difficulty will be experienced in copying the Indore plan. All the urine can be absorbed: all the soiled bedding can be used in the compost pits every morning.

In countries like Great Britain and North America, where labour is both scarce and dear, objection will at once be raised to the Indore plan. Concrete or pitched floors are here the rule. The valuable urine and dung are often removed to the drains by a water spray. In such cases, however, the indispensable urine could either be absorbed on the floors themselves by the addition to the bedding of substances like peat and sawdust mixed with a little earth, or the urine could be directed into small bricked pits just outside the building, filled with any suitable absorbent which is periodically removed and renewed. In this way liquid manure tanks can be avoided. At all costs the urine must be used for composting.

3. *Bases for Neutralizing Excessive Acidity.* In the manufacture of humus the fermenting mixture soon becomes acid in reaction. This acidity must be neutralized, otherwise the work of the micro-organisms cannot proceed at the requisite speed. A base is therefore necessary. Where the carbonates of calcium or potassium are available in the form of powdered chalk or limestone, or wood ashes, these materials either alone, together, or mixed with earth, provide a convenient base for maintaining the general reaction within the optimum range (pH 7·0 to 8·0) needed by the micro-organisms which break down cellulose. Where wood ashes, limestone, or chalk are not available, earth can be used by itself. Slaked lime can also be employed, but it is not so suitable as the carbonate. Quicklime is much too fierce a base.

4. *Water and Air.* Water is needed during the whole of the

period during which humus is being made. Abundant aeration is also essential during the early stages. If too much water is used the aeration of the mass is impeded, the fermentation stops and may soon become anaerobic too soon. If too little water is employed the activities of the micro-organisms slow down and then cease. The ideal condition is for the moisture content of the mass to be maintained at about half saturation during the early stages, as near as possible to the condition of a pressed-out sponge. Simple as all this sounds, it is by no means easy in practice simultaneously to maintain the moisture content and the aeration of a compost heap so that the micro-organisms can carry out their work effectively. The tendency almost everywhere is to get the mass too sodden.

The simplest and most effective method of providing water and oxygen together is whenever possible to use the rainfall—which is a saturated solution of oxygen—and always to keep the fermenting mass open at the beginning so that atmospheric air can enter and the carbon dioxide produced can escape.

After the preliminary fungous stage is completed and the vegetable wastes have broken down sufficiently to be dealt with by bacteria, the synthesis of humus proceeds under anaerobic conditions when no special measures for the aeration of the dense mass are either possible or necessary.

PITS VERSUS HEAPS

Two methods of converting the above wastes into humus are in common use. Pits or heaps can be employed.

Where the fermenting mass is liable to dry out or to cool very rapidly, the manufacture should take place in shallow pits. A considerable saving of water then results. The temperature of the mass tends to remain high and uniform. Sometimes, however, composting in pits is disadvantageous on account of water-logging by storm water, by heavy rain, and by the rise of the ground-water from below. All these result in a wet sodden mass in which an adequate supply of air is out of the question. To obviate such water-logging the composting pits are: (1) surrounded by a catch-drain to cut off surface water; (2) protected by a thatched roof where the rainfall is high and heavy bursts of monsoon rain are the rule; or (3) provided with soakaways at suitable points combined with a slight slope of the floors of the pit towards the drainage corner. Where there is a pronounced rise in the water-table during the rainy season, care must be taken, in siting the pits, that they are so placed that there is no invasion of water from below.

To save the expense of digging pits and to use up sites where excavation is out of the question, composting in heaps is practised. A great deal can be done to increase the efficiency of the heap by protecting the composting area from storm water by means of catch-drains and by suitable shelter from wind, which often prevents all fermentation on the more exposed sides of the heap. In temperate climates heaps should always face the south, and wherever possible should be made in front of a south wall and be protected from wind on the east and west. The effect of heavy rain in slowing down fermentation can be reduced by increasing the size of the heap as much as possible. Large heaps always do better than small ones.

In localities of high monsoon rainfall like Assam and Ceylon, there is a definite tendency to provide the heap or the pit with a grass roof so that the fermentation can proceed at an even rate and so that the annual output is not interfered with by temporary water-logging. After a year or two of service the roof itself is composted. In Great Britain thatched hurdles can be used.

CHARGING THE HEAPS OR PITS

A convenient size for the compost pits (where the annual output is in the neighbourhood of 1,000 tons) is 30 feet by 14 feet and 3 feet deep with sloping sides. The depth is the most important dimension on account of the aeration factor. Air percolates the fermenting mass to a depth of about 18 to 24 inches only, so for a height of 36 inches extra aeration must be provided. This is arranged by means of vertical vents, every 4 feet, made by a light crowbar as each section of the pit is charged.

Charging a pit 30 feet long takes place in six sections each 5 feet wide. The first section, however, is left vacant to allow of the contents being turned. The second section is first charged. A layer of vegetable wastes about 6 inches deep is laid across the pit to a width of 5 feet. This is followed by a layer of soiled bedding or farm-yard manure 2 inches in thickness. The layer of manure is then well sprinkled with a mixture of urine earth and wood ashes or with earth alone, care being taken not to add more than a thin film of about one-eighth of an inch in thickness. If too much is added aeration will be impeded. The sandwich is then watered where necessary with a hose fitted with a rose for breaking up the spray. The charging and watering process is then continued as before until the total height of the section reaches 5 feet. Three vertical aeration vents, about 4 inches in diameter, are then made in the mass by working a crowbar from side to side. The first vent is in the centre, the other two midway between the centre and the sides. As the pit is 14 feet wide and there are three vents, these will be 3 feet 6 inches apart. The next section of the pit (5 feet wide) is then built up close to the first and watered as before. When five sections are completed the pit is filled. The advantages of filling a pit or making a heap in sections to the full height of 5 feet are: (1) fermentation begins at once in each section and no time is lost; (2) no trampling of the mass takes place; (3) aeration vents can be made in each completed section without standing on the mixture.

In dry climates each day's contribution to the pit should again be lightly watered in the evening and the watering repeated the next morning. In this way the first watering at the time of charge is added in three portions—one at the actual time of charging, in the evening after charging is completed and again the next morning after an interval of twelve hours. The object of this procedure is to give the mass the necessary time to absorb the water.

The total amount of water that should be added at the beginning of fermentation depends on the nature of the material, on the climate and on the rainfall. Watering as a rule is unnecessary in Great Britain. If the material contains about a quarter by volume of fresh greenstuff the amount of water needed can be considerably reduced. In rainy weather when everything is on the damp side no water at all is needed. Correct watering is a matter of local circumstances and of individual judgement. At no period should the mass be wet: at no period should the pit be allowed to dry out completely. At the Iceni Nurseries in South Lincolnshire in Great Britain, where the annual rainfall is about

24 inches and a good deal of fresh green market-garden refuse is composted, watering the heaps at all stages is unnecessary. At Indore in Central India where the rainfall was about 50 inches, which fell in about four months, watering was always essential except during the actual rainy season. These two examples prove that no general rule can ever be laid down as to the amount of water to be added in composting. The amount depends on circumstances. The water needed at Indore was from 200 to 300 gallons for each cubic yard of finished humus.

As each section of the pit is completed, everything is ready for the development of an active fungous growth, the first stage in the manufacture of humus. It is essential to initiate this growth as quickly as possible and then to maintain it. As a rule it is well established by the second or third day after charging. Soon after the first appearance of fungous growth the mass begins to shrink and in a few days will just fill the pit, the depth being reduced to about 36 inches.

Two things must be carefully watched for and prevented during the first phase: (1) the establishment of anaerobic conditions caused generally by over-watering or by want of attention to the details of charging; it is at once indicated by smell and by the appearance of flies attempting to breed in the mass; when this occurs the pit should be turned at once; (2) fermentation may slow down for want of water. In such cases the mass should be watered. Experience will soon teach what amount of water is needed at the time of charge.

TURNING THE COMPOST

To ensure uniform mixture and decay and to provide the necessary amount of water and air for the completion of the aerobic phase it is necessary to turn the material twice.

First turn. The first turn should take place between 2 and 3 weeks after charging. The vacant space, about 5 feet wide, at the end of the pit allows the mass to be conveniently turned from one end by means of a pitchfork. The fermenting material is piled up loosely against the vacant end of the pit, care being taken to turn the unaltered layer in contact with the air into the middle of the new heap. As the turning takes place, the mass is watered, if necessary, as at the time of charging, care being taken to make the material moist but not sodden with water. The aim should be to provide the mass with sufficient moisture to carry on the fermentation to the second turn. To achieve this sufficient time must be given for the absorption of water. The best way is to proceed as at the time of charging and 'add any water needed in two stages—as the turning is being done and again next morning. Another series of vertical air vents 3 feet 6 inches apart should be made with a crowbar as the new heap is being made.

Second turn. About five weeks after charge the material is turned a second time but in the reverse direction. By this time the fungous stage will be almost over, the mass will be darkening in colour and the material will be showing marked signs of breaking down. From now onwards bacteria take an increasing share in humus manufacture and the process becomes anaerobic. The second turn is a convenient opportunity for supplying sufficient water for completing the fermentation. This should be added during the actual turning and again the next morning to bring the moisture content to the ideal condition—that of a pressed-out sponge. It will be observed as manufacture proceeds that the mass crumbles and that less and less difficulty occurs in keeping the material moist. This is due to two things: (1) less water is needed in the fermentation; (2) the absorptive and water-holding power of the mass rapidly increase as the stage of finished humus is approached.

Soon after the second turn the ripening process begins. It is during this period that the fixation of atmospheric nitrogen takes place. Under favourable circumstances as much as 25 per cent. of additional free nitrogen may be secured from the atmosphere.

The activity of the various micro-organisms which synthesize humus can most easily be followed from the temperature records. A very high temperature, about 65° C. (149° F.), is established at the outset, which continues with a moderate downward gradient to 30° C. (86° F.) at the end of ninety days. This range fits in well with the optimum temperature conditions required for the micro-organisms which break down cellulose. The aerobic thermophyllic bacteria thrive best between 40° C. (104° F.) and 55° C. (131° F.). Before each turn, a definite slowing down in the fermentation takes place: this is accompanied by a fall in temperature.

E

As soon as the mass is re-made, when more thorough admixture with copious aeration occurs, there is a renewal of activity during which the undecomposed portion of the vegetable matter from the outside of the heap or pit is attacked. This activity is followed by a distinct rise in temperature.

THE STORAGE OF HUMUS

Three months after charge the micro-organisms will have fulfilled their task and humus will have been completely synthesized. It is now ready for the land. If kept in heaps after ripening is completed, a loss in efficiency must be faced. The oxidation processes will continue. Nitrification will begin, resulting in the formation of soluble nitrates. These may be lost either by leaching during heavy rain or they will furnish the anaerobic organisms with just the material they need for their oxygen supply. Such losses do not occur to anything like the same extent when the humus is banked by adding it to the soil. Freshly prepared humus is perhaps the farmer's chief asset and must therefore be looked after as if it were actual money. It is also an important section of the live stock of the farm. Although this live stock can only be seen under the microscope, it requires just as much thought and care as the pigs which can be seen with the naked eye. If humus must be stored it should be kept under cover and turned from time to time.

OUTPUT

The output of compost per annum obviously depends on circumstances. At the Institute of Plant Industry, Indore, where the supply of urine and dung was always greater than that of vegetable waste, fifty cartloads (each 27 c. ft.) of ripe compost, i.e. 1,350 cubic feet or 50 cubic yards, could be prepared from one pair of oxen. Had sufficient vegetable wastes been available the quantity could have been at least doubled. The work-cattle at Indore were of the Malvi breed, about three-quarters the size of the average milking-cow of countries like Great Britain. The urine and dung of an average English cow or bullock, therefore, if properly composted with ample wastes would produce about sixty cartloads of humus a year, equivalent to about 1,600 cubic feet or 60 cubic yards.

As the moisture content of humus varies from 30 to 60 per cent. during the year, it is impossible to record the output in tons unless the percentage of water is determined. The difficulty can be overcome by expressing the output in cubic feet or cubic yards. The rate of application per acre should also be stated as so many cubic feet or cubic yards.

In devising the Indore Process the fullest use was made of agricultural experience including that of the past. After the methods of Nature, as seen in the forest, the practices which throw most light on the preparation of humus are those of the Orient, which have been described by King in *Farmers of Forty Centuries*. In China a nation of observant peasants has worked out for itself simple methods of returning to the soil all the vegetable, animal, and human wastes that are available: a dense population has been maintained without any falling off in fertility.

Coming to the more purely laboratory investigations on the production of humus, two proved of great value in perfecting the Indore Process: (1) the papers of Waksman in which the supreme importance of micro-organisms in the formation of humus was consistently stressed, and (2) the work of H. B. Hutchinson and E. H. Richards on artificial farm-yard manure. Waksman's insistence on the role of micro-organisms in the formation of humus as well as on the paramount importance of the correct composition of the wastes to be converted has done much to lift the subject from a morass of chemical detail and empiricism on to the broad plane of biology to which it rightly belongs. Once it was realized that composting depended on the work of fungi and bacteria, the reform of the various composting systems which are to be found all over the world could be taken in hand. The essence of humus manufacture is first to provide the organisms with the correct raw material and then to ensure that they have suitable working conditions. Hutchinson and Richards come nearest to the Indore Process but two fatal mistakes were made: (1) the use of chemicals instead of urine as an activator in breaking down vegetable wastes, and (2) the patenting of the ADCO process. Urine consists of the drainage of every cell and every gland of the animal body and contains not only the nitrogen and minerals needed by the fungi and bacteria which break down cellulose, but all the accessory growth substances as well. The ADCO powders merely supply factory-made chemicals as well as lime—a far inferior base to the wood ashes and soil used in the Indore Process. It focuses attention on yield rather than on quality. It introduces into composting the same fundamental mistake that is being made in farming, namely, the use of chemicals instead of natural manure. Further, the patenting of a process (even when, as in this case, the patentees derive no personal profit) always places the investigator in bondage; he becomes the slave to his own scheme; rigidity takes the place of flexibility; progress then becomes difficult, or even impossible. The ADCO process was patented in 1916: in 1940 the method to all intents and purposes remains unchanged.

The test of any process for converting the waste products of agriculture into humus is flexibility and adaptability to every possible set of conditions. It should also develop and be capable of absorbing new knowledge and fresh points of view as they arise. Finally, it should be suggestive and indicate new and promising lines of research. If the Indore Process can pass these severe tests it will soon become woven into the fabric of agricultural practice. It will then have achieved permanence and will have fulfilled its purpose—the restitution of their manurial rights to the soils of this planet. In the next four chapters the progress made during the last eight years towards this ideal will be described.

BIBLIOGRAPHY

HOWARD, A., and HOWARD, G. L. C. *The Application of Science to Crop-Production*, Oxford University Press, 1929.
HOWARD, A., and WAD, Y. D. *The Waste Products of Agriculture: Their Utilization as Humus*, Oxford University Press, 1931.

CHAPTER V

PRACTICAL APPLICATIONS OF THE INDORE PROCESS

AFTER the first complete account of the Indore Process was published in 1931, the adoption of the method at a number of centres followed very quickly. The first results were summarized in a lecture which appeared in the issue of the *Journal of the Royal Society of Arts* of December 8th, 1933. About 2,000 extra copies of this lecture were printed and distributed during the next two years. By the end of 1935 it became evident that the method was making very rapid headway all over the world: an increasing stream of interesting results were reported. These were described in a second lecture on November 13th, 1935, which was printed in the *Journal* of the Society of November 22nd, 1935. This lecture was then re-published in pamphlet form. In all 6,425 extra copies of this second lecture have been distributed. During 1936 still further progress was made, a brief account of which appeared in the *Journal of the Royal Society of Arts* of December 18th, 1936; 7,500 copies being printed. Two translations of the 1935 lecture have been published. The first in German in *Der Tropenpflanzer* of February 1936, the second in Spanish in the *Revista del Instituto de Defensa del Café de Costa Rica* of March 1937.

These papers did much to make the Indore Process known all over the world and to start a number of new and active composting centres. The position, as reached by July 1938, was briefly sketched in a paper which was published in the *Journal of the Ministry of Agriculture* of Great Britain of August 1938.

In this and succeeding chapters an attempt will be made to sum up progress to the time of going to press. It will be convenient in the first place to arrange this information under crops.

COFFEE

The first centre in Africa to take up the process was the Kingatori Estate near Kyambu, a few miles from Nairobi, where work began in February 1933. By the purest accident I saw the first beginnings of composting at this estate. This occurred in the course of a tour round Africa which included a visit to the Great Rift Valley. As I was about to start from Nairobi on this expedition, Major Belcher, the Manager of Kingatori, called upon me and said that he had just been instructed by·Major Grogan, the proprietor of the Estate, to start the Indore Process and to convert all possible wastes into humus. He asked me to help him and to discuss various practical details on the spot. I gave up the tour to the Great Rift Valley and spent the day on the Kingatori Estate instead, where it was obvious from the general condition of the bushes and the texture of the soil that a continuous supply of freshly made humus would transform this estate, which I was told was representative of the coffee industry near Nairobi. In a letter dated September 19th, 1933, Major Belcher reported his first results as follows:

'I have 30 pits in regular use now, and am averaging 5 tons of ripe compost from each pit. This will give me a dressing of 3½ tons per acre per annum and should, I think, gradually bring the soil into really good condition.

'I have already dressed 30 acres, but it is a little early to see any result. It is 30 acres of young 4-year old coffee bearing a heavy crop. At the moment it is looking splendid, and if it keeps it up until the crop is picked in December and enables the trees to bear heavily again next year there will be no doubt in my mind that the compost is responsible. Young trees with a big crop are very apt to suffer from die-back of the primaries and light beans and no crop in the following year. There is no sign of this at present.

'I have had many interested visitors, and the Nairobi bookseller has to keep sending for more copies of *The Waste Products of Agriculture*.

'The District Commissioner at Embu has taken up the process extensively with the double purpose of improving village sanitation and the fertility of the soil. In fact, he started it some time before we did.

'I understand that it will shortly be made illegal to export goat and cattle manure from the native reserves, in which case your process will be taken up by most of the European farmers in the Colony. One very influential member of the coffee industry remarked to me that he thought your process would revolutionize coffee-growing and another said that he considered it was the biggest step forward made in the last ten years.'

Two years later he sent me a second report in which he stated that during the last 28 months 1,660 tons of compost, containing about 1·5 per cent. of nitrogen, had been manufactured on this estate and applied to the land. The cost per ton was 4s. 4d.— chiefly the expense involved in collecting raw material. The work in progress had been shown to a constant stream of visitors from other parts of Kenya, the Rhodesias, Uganda, Tanganyika, and the Belgian Congo. Major Belcher has lost count of the actual numbers.

This pioneering work has done much more than weld the Indore Process into the routine work of the estate. It has served the purpose of an experiment station and a demonstration area for the coffee industry throughout the world. Many new centres followed Kingatori. The rapid spread of the method is summed up by Major Grogan in a letter dated Nairobi, May 15th, 1935, as follows:

'You will be glad to know that your process is spreading rapidly in these parts and has now become recognized routine practice on most of the well conducted coffee plantations. The cumulative effect of two years on my plantation is wonderful. I have now established all round my pits a large area of elephant grass for the purpose of providing bulk, and we have made quite a lot of pocket-money by selling elephant grass cuttings to the country-side. I am now searching for the best indigenous legumes to grow in conjunction with the elephant grass and am getting very hopeful results from the various *Crotalarias* and *Tephrosias* which I have brought up from the desert areas of Taveta. They get away quickly and so hold their own against the local weeds.

Major Grogan in referring to the spread of the Indore Process in East Africa, has omitted one very material factor, namely, his personal share in this result. He initiated the earliest trial on the Kingatori Estate and has always insisted on the method having a square deal in Kenya. In Tanganyika the influence of Sir Milsom Rees, G.C.V.O., has led to similar results.

This example of the introduction and spread of the Indore Process on the coffee estates of Kenya and Tanganyika has been given in detail for three reasons: (1) it was one of the earliest applications of the Indore method to the plantation industries; (2) it is typical of many other similar applications elsewhere; and (3) it first suggested to me a new field of work during retirement in which the research experience of a lifetime could be fully utilized.

Kenya and Tanganyika are only two of the coffee centres of the world. The largest producer is the New World. Here satisfactory progress has been made following the publication in the *West India Committee Circular* of April 23rd, 1936, of a short account of the Indore Process. This led to important developments, first in Costa Rica and then in Central and South America, as a result of a Spanish translation by Señor Don Mariano Montealegre of my 1935 lecture to the Royal Society of Arts to which reference has already been made. This was widely read in all parts of Latin America: the lecture drew attention to the vital necessity of organic matter in the production of coffee in the New World. During the next two years no less than seven Spanish translations of my papers on humus were published in the *Revista del Instituto de Defensa del Café de Costa Rica*. In January 1939 a special issue of the *Revista* entitled *En Busca del Humus* (In Quest of Humus) appeared. This was devoted to a collection of papers describing the Indore Process and the various developments of the last eight years.

The marked response of coffee to humus in Africa, India, and the New World suggested that the crop would prove to be a mycorrhiza-former. A number of samples of the surface roots of coffee plants were duly collected in Travancore, Tanganyika, and Costa Rica and sent home for examination. In all cases they showed the mycorrhizal association.

TEA

The East African results with coffee naturally suggested that something should be done with regard to tea—a highly organized plantation industry with the majority of the estates arranged in large groups, controlled by a small London Directorate largely recruited from the industry itself. The problem was how best to approach such an organization. In 1934 my knowledge of tea and of the tea industry was of the slightest: I had never grown a tea plant, let alone managed a tea plantation. I had only visited two tea estates, one near Nuwara Eliya in Ceylon in 1908 and the other near Dehra Dun in 1918. I had, however, kept in touch with the research work on tea. While I was debating this question Providence came to my assistance in the shape of a request from a mutual friend to help Dr. C. R. Harler (who had just been retrenched when the Tocklai Research Station, maintained by the Indian Tea Association, was reorganized in 1933) to find a new and better opening, if possible one with more scope for independent and original work. I renewed my acquaintance with Dr. Harler and suggested he should take up the conversion into humus of the waste products of tea estates. He was very interested and shortly afterwards (August 1933) accepted the post of Scientific Officer to the Kanan Devan Hills Produce Co. in the High Range, Travancore, which was offered him by Messrs. James Finlay & Co., Ltd. On taking up his duties in this well-managed and highly efficient undertaking, Dr. Harler secured the active interest of the then General Manager, Mr. T. Wallace, and set to work to try out the Indore Process on an estate scale at his head-quarters at Nullatanni, near Munnar. No difficulties were met with in working the method: ample supplies of vegetable wastes and cattle manure were available: the local labour took to the work and the Estate Managers soon became enthusiastic.

On receipt of this information I made inquiries from Dr. H. H. Mann, a former Chief Scientific Officer of the Indian Tea Association, as to whether the live wires among the London Directorate of the tea industry included anybody likely to be particularly interested in the humus question.

I was advised to see Mr. James Insch, one of the Managing Directors of Messrs. Walter Duncan & Co. At Mr. Insch's request an illustrated paper of instructions for the use of the Managers of the Duncan Group was drawn up in October 1934 and 250 copies were printed. The Directors of other groups of tea estates soon began to consider the Indore Process and 4,000 further copies of the paper of instructions were distributed. By the end of 1934 fifty-three estates of the Duncan Group in Sylhet, Cachar, the Assam Valley, the Dooars, Terai, and the Darjeeling District had made and distributed sample lots of humus, about 2,000 tons in all. At the time of writing, December 1939, the estates of the Duncan Group alone expect to make over 150,000 tons of humus a year. Similar developments have occurred in a number of other groups notably on the estates controlled by Messrs. James Finlay & Co., who have never lost the lead in manufacturing humus which naturally followed from the pioneering work done by Dr.

Harler in Travancore. A good beginning has been made. The two strongest groups of tea estates in the East have become compost-minded.

It is exceedingly difficult to say exactly how much humus is being made at the present time on the tea estates of the British Empire. It is possible only to give a very approximate figure. In April 1938 Messrs. Masefield and Insch stated: 'It is probably no exaggeration to say that to-day a million tons of compost are being made annually on the tea estates of India and Ceylon, and this has been accomplished within a period of 5 years.' Since this was written the tea estates of Nyasaland and Kenya have also taken up the Indore Process with marked success.

These developments have been accompanied by a considerable amount of discussion. Two views have been and are still being held on the best way of manuring tea. One school of thought, which includes the tea research institutes, considers that as the yield of leaf is directly influenced by the supply of combined nitrogen in the soil, the problem of soil fertility is so simple as to reduce itself to the use of the cheapest form of artificial manure—in this case sulphate of ammonia. This view is naturally vigorously supported by the artificial manure interests. The results obtained with sulphate of ammonia on small plots at Tocklai and Borbhetta are triumphantly brought forward to clinch the argument which amounts to this: that tea can be grown on a conveyor-belt lubricated by chemical fertilizers. The weaknesses of such an argument are obvious. These small plots do not represent anything in the tea industry: they only represent themselves. It is impossible to run a small plot or to manufacture and sell its produce as a tea-garden is conducted. In other words the small plot is not practical politics. Again, land like Tocklai and Borbhetta which responds so markedly to sulphate of ammonia must be badly farmed, otherwise artificials would not prove so potent. The tendency all the world over is that as the soil becomes more fertile artificials produce less and less result until the effect passes off altogether. Bad farming and an experimental technique which will not hold water are poor foundations on which to found a policy. The use of replicated and randomized plots, followed by the higher mathematics in interpreting the results of these small patches of land, can do nothing to repair the fundamental unsoundness of the Tocklai procedure. It stands self-condemned. Further, the advocates of sulphate of ammonia for the tea plant seem to have forgotten that a part at least of the extra yield obtained with this manure may be due to an increase in soil acidity. Tea, as is well known, needs an acid soil: sulphate of ammonia increases acidity.

The humus school of thought takes the view that what matters in tea is quality and a reserve of soil fertility such as that created by the primeval forest: that this can only be obtained by freshly prepared humus made from vegetable *and* animal wastes and by the correct use of shade trees, green-manure crops, and the prevention of soil erosion. The moment the tea soils can be made really fertile, the supply of nitrogen to the plant will take care of itself and there will be no need to waste money in securing the fleeting benefits conferred by artificials. The problem therefore of the manuring of tea is not so much the effect of some dressing on the year's yield but the building up of a store of fertility. In this way the manurial problem and the stability of the enterprise as a going concern become merged into one. It is impossible to separate the profit and loss account and the balance-sheet of a composting programme because the annual dressings of humus influence both.

It will be interesting to watch the results of this struggle in a great plantation industry. At the moment a few of the strongest and most successful groups are taking up humus and spend little or nothing on artificials. Other companies, on the other hand, are equally convinced that their salvation lies in the use of cheap chemical fertilizers. Between these two extremes a middle course is being followed—humus supplemented by artificials. Mother earth, rather than the advocates of these various views, will in due course deliver her verdict.

Can the tea plant itself throw any light on this controversy or is it condemned to play a merely passive role in such a contest? Has the tea bush anything to say about its own preference? If it has, its representations must at the very least be carefully considered. The plant or the animal will answer most queries about its needs if the question is properly posed and if its response is carefully studied.

During the early trials of the Indore Process it became apparent that the tea plant had something very interesting to communicate on the humus question. Example after example came to my notice where such small applications of compost as five tons to the acre were at once followed by a marked improvement in growth, in general vigour and in resistance to disease. Although very gratifying, in one sense these results were somewhat disconcerting. If humus acts only indirectly by increasing the fertility of the soil, time will be needed for the various physical, biological, and chemical changes to take place. If the plant responds at once, some other factor besides an improvement in fertility must be at work. What could this factor be?

In a circular letter issued on October 7th, 1937, to correspondents in the tea industry, I suggested that the most obvious explanation of any sudden improvement in tea, observed after one application of compost, is the effect of humus in stimulating the mycorrhizal relationship which is known to occur in the roots of this crop.

In the course of a recent tour (November 1937 to February 1938) to tea estates in the East, I examined the root system of a number of tea plants which had been manured with properly made compost, and found everywhere the same thing—numerous tufts of healthy-looking roots associated with rapidly developing foliage and twigs much above the average. Both below and above ground humus was clearly leading to a marked condition of well-being. When the characteristic tufts of young roots were examined microscopically, the cortical cells were seen to be literally overrun with mycelium and to a much greater extent than is the rule in a really serious infection by a parasitic fungus. Clearly the mycorrhizal relationship was involved. These necessarily hasty and imperfect observations, made in the field, were soon confirmed and extended by Dr. M. C. Rayner and Dr. Ida Levisohn, who examined a large number of my samples including a few in which artificials only were used, or where the soils were completely exhausted and the garden had become derelict with perhaps only half the full complement of plants. In these cases the characteristic tufts of healthy roots were not observed; root development and growth were both defective; the mycorrhizal relationship was either absent or poorly developed. Where artificials were used on worn-out tea, infection by brownish hyphae of a *Rhizoctonia*-like fungus (often associated with mild parasitism) was noticed. Whenever the roots of tea, manured with properly made compost, were critically examined, the whole of the cortical tissues of the young roots always showed abundant endotrophic mycorrhizal infection, the mainly intracellular mycelium apparently belonging to one fungus. The fungus was always confined to the young roots and no extension of the infection to old roots was observed. In the invaded cells the mycelium exhibits a regular cycle of changes from invasion to the clumping of the hyphae around the cell nuclei, digestion and disintegration of their granular contents, and the final disappearance of the products from the cells.

Humus in the soil therefore affects the tea plant direct by means of a middleman—the mycorrhizal relationship. Nature has provided an interesting piece of living machinery for joining up a fertile soil with the plant. Obviously we must pay the closest attention to the response—as regards yield, quality, and disease resistance—which follows the use of this wonderful bit of mechanism. We must also see that the humus content of the soil is such that the plant can make the fullest possible use of its own machinery.

The mycorrhizal relationship in tea and its obvious bearing on the nutrition of the plant places the manurial problems of this crop on a new plane—that of applied biology. The well-being of the tea plant does not depend on the cheapest form of nitrogen but on humus and the consequences of the mycorrhizal relationship. We are obviously dealing with a forest plant which thrives best on living humus—not on the dead by-product of a factory.

It is easy to test the correctness of this view. It can be done in two ways: (1) by a comparison of tea seedlings grown on sub-soil (from which the surface soil containing humus has been removed) and manured either with a complete artificial mixture or with freshly prepared humus, and (2) by observing the effect of artificials on a tea-garden where the soil is really fertile. Such trials have already been started. In the case of seedlings grown on sub-soil manured with: (1) 20 tons of humus to the acre, or (2) the equivalent amount of NPK in the form of artificials, Mr. Kenneth Morford has obtained some very interesting results at Mount Vernon in Ceylon. Nine months after sowing, the humus plot was by far the better—the plants were 10 inches high, branched, with abundant, healthy, dark green foliage. The plots with artificials were 6 inches high, unbranched, with sparse, unhealthy, pale foliage. An examination of the root systems was illuminating. The humus plants developed a strong tap root 12 inches long; the artificials plot showed little attempt to develop any tap root at all, only extensive feeding roots near the surface. The root system at once explained why the humus plot resisted drought and why the artificial manure plot was so dependent on watering. Mr. Morford's experiment should be repeated in some of the other tea areas of the East. The results will speak for themselves and will need no argument.

The effect of sulphate of ammonia on a really fertile soil is most interesting. As would be expected the results have been for the most part almost negative, because there is no limiting factor in the shape of a deficiency of nitrogen, phosphorus, or potash under such conditions. On old estates, where organic matter has not been regularly replaced, resulting in the loss of much of the original fertility, such an experiment would give a clear indication as to whether, under existing management, the soil is losing, maintaining, or gaining in fertility. Given an adequate supply of humus in the soil, the mycorrhizal relationship and the nitrification of organic matter, when allowed to work at top speed, are all that the plant needs to produce a full crop of the highest quality possible under local conditions. The tea plant therefore is already preparing its own evidence in the suit—Humus versus Sulphate of Ammonia.

The problem of the manuring of tea is straightforward. It consists in converting the mixed vegetable wastes of a tea estate and of the surrounding land into humus by means of the urine and dung of an adequate herd of live stock—cattle, pigs, or goats. As the tea districts are situated in regions of high rainfall it will be necessary in many cases to protect the heaps or pits from heavy rain. Ample vegetable waste must also be provided. The solution of the practical problems involved will necessarily depend on local conditions. At Gandrapara in the Dooars, an estate influenced by the south-west monsoon, Mr. J. C. Watson has set about the provision of an ample supply of humus in a very thorough-going manner, an account of which will be found in Appendix A (p. 225). It cannot fail to interest not only the producers of tea but the whole of the plantation industries as well.

The conversion of vegetable and animal wastes into humus is only one aspect of the soil-fertility problem of a tea-garden. There are a number of others such as the use of shade trees, drainage, prevention of erosion, the best manner of utilizing tea prunings and green-manure, the utilization of water-weeds like the water hyacinth, the treatment of root disease, the raising of seed, the manufacture of humus from vegetable wastes only, and the effect of artificial manures on the quality of tea. These will now be briefly discussed.

Generally speaking, more attention is paid to shade trees in North-East India than in South India and Ceylon. There is a tendency for shade to decrease as one proceeds south. It may be that the factor which has determined the invariable use of shade in North-East India is the intense dryness and heat of the period March to June which does not occur in the south. As, however, tea is a forest plant and tea-growing must always be looked upon as applied forestry, it would seem to be a mistake to reduce shade too much. The organic matter provided by the roots and leaves of the shade trees, the protection they afford the soil from the sun, wind, and rain, and the well-known advantage of mixed cropping must all be very important factors in the maintenance of fertility. This is borne out by the superior appearance of the tea on well-shaded estates in Ceylon compared with that on land alongside where the shade trees have been removed.

A large amount of the vegetable wastes on a tea estate consists of prunings and green-manure plants. These are either forked in, buried in long shallow trenches, or made into humus. Is there any more effective method of dealing with these wastes? When the tea is pruned the plant makes a new bush. Could it not be induced to re-make a portion of its root system at the same time in well-aerated rich soil? I think it could. On estates provided with adequate shade and contour drains, the following two methods of composting tea prunings and green-manure might be tried out:

1. This material should be forked in with a dressing of compost

at the rate of 5 to 10 tons an acre. Decay will then be much more rapid and effective than is now the case. This method of the sheet composting of tea prunings has been tried out and found successful at Gandrapara (p. 227).

2. The prunings and green-manure should be composted in small pits between alternate rows of tea. The pits should be 2 feet long and 1½ feet wide and 9 inches to a foot deep, parallel to the drains or contour drains and so arranged that the roots of every tea plant come in contact with one pit only. The pits are then nearly half filled with mixed tea and green-manure prunings, which are then covered with a thin layer of compost or cattle manure. More green material is added until the pit is nearly full. It is then covered with three inches of soil. The pits now become small composting chambers; humus is produced while the tea is not growing a crop; earthworms are encouraged; the roots of the neighbouring tea plants soon invade the pit; a portion of the root system of all the tea plants of the area is then re-created in highly fertile, permeable soil. When the pruned bushes need tipping on estates where the first picking is not manufactured, another set of similar pits can be made in the vacant spaces between the first pits in each line and similarly filled.

When next the bushes are pruned exactly the same procedure can be carried out in the hitherto undisturbed spaces between the rows of tea.

When the fourth set of pits has been made each tea bush will have completed the re-creation of a large portion of its root system in rich earth.

The first large-scale trial of the pit method was begun at Mount Vernon in Ceylon in January 1938. The results have been satisfactory in all respects: the yield of tea has increased: the plants have resisted drought: the cost of the work has proved to be a sound investment.

On several tea estates in Assam the low-lying areas among the tea are used for the growth of water hyacinth for the compost heaps. When this material forms a quarter to a third of the volume of the heap, watering during the dry season can be reduced very considerably. About three-quarters of the weed is harvested, the remainder being left to produce the next year's crop. As water hyacinth is known to diminish the number of mosquitoes it might pay a tea estate from the point of view of malaria control only to grow this plant for composting on all low-lying areas. When water hyacinth becomes widely cultivated on the tea estates for humus manufacture, the labour employed will undoubtedly carry the news to the great rice areas of North-East India. Here one of the greatest advances in food production in the world can be achieved by the conversion of water hyacinth first into humus and then into rice.

In many of the tracts which produce tea small areas occur in which the bushes are attacked by root disease. It is probable that local soil pans, some distance below the surface, are holding up the drainage and that this stagnant water lowers the natural resistance of the tea plant. I suggested in the Report on my tour that vertical pillar drains, filled with stones, pebbles, or even surface soil, might prevent these troubles. Similar drains are used in Sweden with good results.

The weakest link in the tea industry is the production of seed. During the whole of my tour I saw few really well-managed seed gardens. It is essential that the trees which bear seed should be properly selected, adequately spaced, well drained, and manured with freshly prepared compost. Nature will provide an automatic method of seed control. If diseases appear on the trees or in the seeds something is wrong. Only if the trees and seed are healthy, vigorous, and free from pests, is the produce of such trees fit for raising plants, which in China are said to last a hundred years. The tea plant must have a good start in life.

In Ceylon particularly, attempts have been made to prepare humus without animal wastes. The results have not fulfilled expectation. The breaking down of such resistant material as the leaves and prunings of tea is then unsatisfactory: the organisms which synthesize humus are not properly fed: the residues of these organisms which form an important part of the final humus lack the contributions of the animal. No one has yet succeeded in establishing an efficient and permanent system of agriculture without live stock. There is no reason therefore to suppose that the tea industry will prove an exception to what, after all, is a rule in Nature.

F

One of the most discussed topics in tea is the effect of artificial manures on quality. The view is widely held that there has been a gradual loss in quality since chemical manures were introduced. One of the planters in the Darjeeling District, Mr. G. W. O'Brien, the proprietor of the Goomtee and Jungpana Tea Estates, who continues to produce tea of the highest quality, informed me in 1935 that he had never used artificials since the estates came under his management thirty-one years ago. The only manure used is cattle manure and vegetable wastes—in other words, humus. The role of the mycorrhizal relationship in tea helps to provide a scientific explanation of these results. There can be little doubt that this relationship will be found to influence the quality of tea as well as the productivity and health of the bush. Humus and the mycorrhizal relationship cannot of course create quality where it never existed: the utmost these factors can achieve is to restore that degree of quality which any locality possessed when first it was brought from forest under tea.

SUGAR-CANE

The waste products of the sugar-cane vary considerably. In peasant agriculture where the whole of the megass is burnt for evaporating the juice in open pans, the chief waste is old cane leaves, cane stumps, and the ashes left by the fuel. On the sugar estates, a number of factory wastes must be added to the above list—filter press cake, some unburnt megass, and the distillery effluent left after the manufacture of alcohol (known in Natal as dunder). The main waste in both cases, however, is the old dead leaves (cane trash), a very difficult material to turn into humus on account of its structure and its chemical composition.

Before the advent of artificial fertilizers, it was the custom on sugar estates to maintain animals—mules and oxen—for transport and for cultivation. These animals were bedded down with cane trash, and a rough farm-yard manure—known in the West Indies as pen manure—was obtained with the help of their wastes. Soon after the introduction of artificial manures, the value of this product began to be assessed on the basis of chemical analysis. Comparisons were made between the cost of production of its content of NPK and that of an equivalent amount of these chemical elements in the form of artificial manure. The result was chemicals soon began to displace pen manure: the animal came to be regarded as an expensive luxury. The advent of the tractor and the motor-lorry settled the question. Why keep expensive animals like mules and oxen which have to be fed from the land when their work can be done more cheaply by machines and imported fuel? The decision to give up animals and farm-yard manure altogether naturally followed because the clearest possible evidence—that of the profit and loss account—was available. Such false reasoning is, alas, only too common in agriculture.

The reaction of the sugar-cane crop itself to this change in manuring was interesting. Two things happened: (1) insect and fungous diseases increased; (2) the varieties of cane showed a marked tendency to run out. These difficulties were met by a constant stream of new seedling varieties. In contrast to this behaviour of the cane on the large estates is that of the same crop grown by the cultivators of northern India where the only manure used is cattle manure and where there is practically no disease and no running out of varieties. The indigenous canes of the United Provinces have been grown for twenty centuries without any help from mycologists, entomologists, or plant breeders.

Why does a variety of cane run out and why does it fall a prey to disease? Sugar-cane is propagated vegetatively from cuttings. When the buds from which the new canes arise are grown with natural manure in India, the variety to all intents and purposes is permanent. On the sugar estates, however, when the buds are raised with chemicals the variety is short-lived. There must be some simple explanation of this difference in behaviour.

What happened in the early days of the sugar estates before the advent of chemicals and before new seedling canes were discovered? In the West Indies, for example, until the last decade of the last century the Bourbon variety was practically the only kind grown. There was little or no disease and this old variety showed no tendency to run out. The experience of the cultivators of the United Provinces of India has therefore been repeated on the estates themselves.

The simplest explanation of the breakdown of cane varieties is that artificials do not really suit the cane and that they lead to incipient malnutrition. If this is so the synthesis of carbohydrates and proteins will be slightly imperfect: each generation of the cane will start somewhat below par. The process will eventually end in a cane with a distinct loss in vegetative vigour and unable to resist the onslaughts of the parasite. In other words, the variety will have run out.

This hypothesis will be transformed into something approaching a principle if it can be proved that the cane is a mycorrhiza-former and is nourished in two ways: (1) by the carbohydrates and proteins synthesized in the green leaves. and (2) by the direct digestion of fungous mycelium in the roots.

Steps were taken during 1938 and 1939 to have the roots of sugar-cane examined in order to test this point of view. Material was obtained from India, Louisiana, and Natal. In all cases the roots exhibited the mycorrhizal association. The large amount of material sent from Natal included canes grown with artificials only, with humus only, and with both. The results were illuminating. Humus is followed by the establishment of abundant mycorrhiza and the rapid digestion of the fungus by the roots of the cane. Artificials tend either to eliminate the association altogether or to prevent the digestion of the fungus by the roots of the cane. These results suggest that the change over from pen manure to artificials is at the root of the diseases of the cane and is the cause of the running out of the variety. We are dealing with the consequences of incipient malnutrition—a condition now becoming very general all over the world in many other crops besides sugar-cane.

These observations leave little doubt that the future policy in cane-growing must be the conversion of cane trash and other wastes into humus. The difficulty in composting cane trash, however, is to start the fermentation and then to maintain it. The leaves are armour-plated and do not easily absorb water. Further, the material is low in nitrogen (about 0·25 per cent.) while the ash (7·3 per cent. of the mass) contains 62 per cent. of silica. The micro-organisms which manufacture humus find it difficult to start on such refractory material. The problem is how best to help them in their work: (1) by getting the trash to absorb water, and (2) by providing them with as much easily fermentable vegetable matter as possible. Molasses where available can be used to help the fermentation. If humus of the highest quality is to be synthesized an adequate supply of urine and dung must also be provided, otherwise a product without the accessory growth substances will result. Given a reasonable supply of urine and

TABLE 2

Composting cane trash in Natal

Composted with	Mois-ture	Loss on ignition	N	Total P_2O_5	Avail-able P_2O_5	Total K_2O	Avail-able K_2O
1. Kraal manure . .	60·5	30·6	0·74	0·28	0·14	S.T.	..
2. Filter cake . .	74·2	44·0	0·67	0·68	0·52	S.T.	..
3. Kraal manure and filter cake . .	61·0	33·3	0·71	0·40	0·28	S.T.	..
4. Kraal manure, filter cake, and molasses	64·8	34·6	0·70	0·40	0·20	T.	S.T.
5. Dunder . . .	28·5	20·0	0·72	0·40	0·21	0·52	0·30
6. Kraal manure, filter cake, ammonium sulphate, and potassium sulphate .	59·2	27·8	1·00	0·42	0·29	0·72	0·49
7. Farm composts with available materials	55·5	27·6	0·78	0·32	0·24	S.T.	..
8. Farm composts with available materials	52·2	29·6	0·67	0·89	0·56	S.T.	..
9. Farm composts with available materials	57·8	33·1	0·91	0·56	0·44	S.T.	..
10. Farm composts with available materials	41·0	30·0	0·84	0·44	0·36	S.T.	..
11. Farm composts with available materials	29·2	9·9	0·67	0·27	0·20	S.T.	..

dung and sufficient easily fermentable vegetable wastes like green-manure, there is no reason why cane trash and the other wastes of a sugar plantation cannot be made into first-class humus and why a sugar estate should not be made to manure itself. The conditions which must be fulfilled are clear from the work already done. Dymond has shown that before composting, cane trash must be allowed to weather a little: the weathered leaves must then be kept moist from the start. In this way the fungi and bacteria are greatly assisted. Filter press cake, dunder, and other wastes all help in the process of conversion, as will be seen from the results of his various experiments carried out in 1938 in Natal (Table 2).

These results are similar to and confirm those obtained by Tambe and Wad at Indore in 1935. In Natal it is estimated that 100 tons of stripped cane will yield about 40 tons of compost containing about 280 lb. of nitrogen and 160 lb. of phosphoric acid.

The main difficulty in composting cane trash must always be the correction of its wide carbon : nitrogen ratio. The problem is a practical one—how best to bring the various wastes together in the cheapest way and then distribute the finished humus to the land. Obviously there can be no hard-and-fast procedure. The correct solution of the problem will vary with the locality: the work is such that it can only be done by the man on the spot.

The sugar estates of the future will in all probability gradually become self-supporting as regards manure. After a time no money will be spent on artificials. The change over from present methods of manuring will, however, take time, and at first a sufficient volume of high-quality humus will be out of the question because the animals maintained will be too few.

What is the best way of using the small amount of humus that can be made at the beginning? This is a very important matter. I suggest that it should be devoted to the land on which the plant material is grown. These canes should be raised in trenches on the Shahjahanpur principle (see Chapter XIV) and every care should be taken to maintain the aeration of the soil during the whole life of the crop. The trenches should be well cultivated and manured with freshly prepared humus, at least three months before planting. These canes should be regarded as the most important on the estate, and no pains should be spared to produce the best possible material. Whether or not immature cane should continue to be planted is a question for the future. What is certain is that cane to be planted should be really well grown in a soil rich in freshly prepared humus. Each crop must start properly. As the supply of organic matter increases on the sugar estates the methods found to give the best results in growing these canes can be extended to the whole estate.

That the above is possible is clear from a study of the work that has been done in India and Natal. In March 1938 Dymond concluded a careful survey of the whole problem in the following words:

'Artificials are easy of application, easily purchased in good times, or not bought at all when times are bad; they form a never-ending topic of conversation with one's neighbours, a source of argument with the vendors; they are a duty and a sop to one's conscience; whereas humus means more labour, more attention, transport and trouble. Nevertheless, humus is the basis of permanent agriculture, artificials the policy of the here to-day and gone to-morrow.'

COTTON

Before taking up research on cotton at the newly founded Institute of Plant Industry at Indore in 1924, a survey of cotton growing in the various parts of India was undertaken. At the same time the research work in progress on this crop was critically examined.

The two most important cotton-growing areas in India are: (1) the black cotton soils of the Peninsula, which are derived from the basalt; (2) the alluvium of North-West India, the deposits left by the rivers of the Indo-Gangetic plain.

In the former there are thousands of examples which indicate beyond all doubt the direction research on this crop should take. All round the villages on the black cotton soils, zones of highly manured land rich in organic matter occur. Here cotton does well no matter the season: the plants are well grown and free from pests: the yield of seed cotton is high. On the similar but un-manured land alongside the growth is comparatively poor: only in years of well distributed rainfall is the yield satisfactory. The limiting factor in growth is the development, soon after the rains set in, of a colloidal condition which interferes with aeration and impedes percolation. This occurs on all black soils, but organic matter mitigates the condition.

On the alluvium of North-West India, a similar limiting factor occurs. Here cotton is grown on irrigation, which first causes the fine soil particles to pack and later on to form colloids. In due course the American varieties in particular show by their growth that they are not quite at home. The anthers often fail to function properly, the plants are unable to set a full crop of seed, the ripening period is unduly prolonged, and the fibre lacks strength, quality, and life. The cause of this trouble is again poor soil aeration, which appears in these soils to lead to a very mild alkali condition. This, in turn, prevents the cotton crop from absorbing sufficient water from the soil. One of the easiest methods of preventing this packing is by assisting the soil to form compound particles with the help of dressings of humus.

The basis of research work on cotton in India was therefore disclosed by a study in the field of the crop itself. The problem was how best to maintain soil aeration and percolation. This could be solved if more humus could be obtained. Good farming methods therefore provided the key to the cotton problems of India.

A study of the research work which has been done all over the world did nothing to modify this opinion. The fundamental weakness in cotton investigations appeared to be the fragmentation of the factors, a loss of direction, failure to define the problems to be investigated, and a scientific approach on far too narrow a front without that balance and stability provided by adequate farming experience.

Steps were therefore taken to accelerate the work on the manufacture of humus which had been begun at the Pusa Research Institute. The Indore Process was the result. It was first necessary to try it out on the cotton crop. The results are summed up in the following Table:

TABLE 3

The increase in general fertility at Indore

Year	Area in acres of improved land under cotton	Average yield in lb. per acre	Yield of the best plot of the year in lb. per acre	Rainfall in inches
1927	20·60	340	384	27·79 (distribution good)
1928	6·64	510	515	40·98 (a year of excessive rainfall)
1929	36·98	578	752	23·11 (distribution poor)

The figures show that, no matter what the amount and distribution of rainfall were, the application of humus soon trebled the average yield of seed cotton—200 lb. per acre—obtained by the cultivators on similar land in the neighbourhood.

In preparing humus at Indore one of the chief wastes was the old stalks of cotton. Before these could be composted they had to be broken up. This was accomplished by laying them on the estate roads, where they were soon reduced by the traffic to a suitable condition for use as bedding for the work-cattle prior to fermentation in the compost pits.

The first cotton grower to apply the Indore Process was Colonel (now Sir Edward) Hearle Cole, C.B., C.M.G., at the Coleyana Estate in the Montgomery District of the Punjab, where a compost factory on the lines of the one at the Institute of Plant Industry at Indore was established in June 1932. At this centre all available wastes have been regularly composted since the beginning: the output is now many thousands of tons of finished humus a year. The cotton crop has distinctly benefited by the regular dressings of humus; the quality of the fibre has improved; higher prices are being obtained; the irrigation water required is now one-third less than it used to be. The neighbouring estates have all adopted composting: many interested visitors have seen the work in progress. One advantage to the Punjab of this work has, however, escaped attention, namely, the importance of the large quantities of well-grown seed, raised on fertile soil, contributed by this estate to the seed distribution schemes of the Provincial Agricultural Department. Plant breeding to be successful involves two things— an improved variety plus seed for distribution grown on soil rich in humus.

The first member of an Agricultural Department to adopt the Indore Method of composting for cotton was Mr. W. J. Jenkins, C.I.E., when Chief Agricultural Officer in Sind, who proved that humus is of the greatest value in keeping the alkali condition in check, in maintaining the health of the cotton plant, and in increasing the yield of fibre. At Sakrand, for example, no less than 1,250 cartloads of finished humus were prepared in 1934–5 from waste material such as cotton stalks and crop residues.

During recent years the Indore Process has been tried out on some of the cotton farms in Africa belonging to the Empire Cotton Growing Corporation. In Rhodesia, for example, interesting results have been obtained by Mr. J. E. Peat at Gatooma. These were published in the *Rhodesia Herald* of August 17th, 1939. Compost markedly improved the fibre and increased the yield not only of cotton but also of the rotational crop of maize.

Why cotton reacts so markedly to humus has only just been discovered. The story is an interesting one, which must be placed on record. In July 1938 I published a paper in the *Empire Cotton Growing Review* (vol. xv, no. 3, 1938, p. 186) in which the role of the mycorrhizal relationship in the transmission of disease resistance from a fertile soil to the plant was discussed. In the last paragraph of this paper the suggestion was made that mycorrhiza 'is almost certain to prove of importance to cotton and the great differences observed in Cambodia cotton in India, in yield as well as in the length of the fibre, when grown on (1) garden land (rich in humus) and (2) ordinary unmanured land, might well be explained by this factor'. In the following number of this *Journal* (vol. xv, no. 4, 1938, p. 310) I put forward evidence which proved that cotton is a mycorrhiza-former. The significance of this factor to the cotton industry was emphasized in the following words:

'As regards cotton production, experience in other crops, whose roots show the mycorrhizal relationship, points very clearly to what will be necessary. More attention will have to be paid to the well tried methods of good farming and to the restoration of soil fertility by means of humus prepared from vegetable *and* animal wastes. An equilibrium between the soil, the plant and the animal can then be established and maintained. On any particular area under cotton, a fairly definite ratio between the number of live stock and the acreage of cotton will be essential. Once this is secured there will be a marked improvement in the yield, in the quality of the fibre and in the general health of the crop. All this is necessary if the mycorrhizal relationship is to act and if Nature's channels of sustenance between the soil and the plant are to function. Any attempt to side-track this mechanism is certain to fail.

'The research work on cotton of to-morrow will have to start from a new base line—soil fertility. In the transition between the research of to-day and that of the future, a number of problems now under investigation will either disappear altogether or take on an entirely new complexion. A fertile soil will enable the plant to carry out the synthesis of proteins and carbohydrates in the green leaf to perfection. In consequence the toll now taken by fungous, insect and other diseases will at first shrink in volume and then be reduced to its normal insignificance. We shall also hear less about soil erosion in places like Nyasaland where cotton is grown, because a fertile soil will be able to drink in the rainfall and so prevent this trouble at the source.'

Confirmation of these pioneering results soon followed. In the *Transactions of the British Mycological Society* (vol. xxii, 1939, p. 274)

Butler mentions the occurrence of mycorrhiza as luxuriantly developed in cotton from the Sudan and also in cotton from the black soils of Gujerat (India). In the issue of *Nature* of July 1st, 1939, Younis Sabet recorded the mycorrhizal relationship in Egypt. In the *Empire Cotton Growing Review* of July 1939 Dr. Rayner confirmed the existence of mycorrhiza in both Cambodia and Malvi cotton grown at my suggestion by Mr. Y. D. Wad at Indore, Central India, in both black cotton soil and in sandy soil from Rajputana.

SISAL

As is well known, the leaves of the sisal plant yield about 93 per cent. of waste material and about 7 per cent. of fibre, of which not more than 5 per cent. is ordinarily extracted. The wastes are removed from the decorticators by a stream of water, usually to some neighbouring ravine or hollow in which they accumulate. Sometimes they are led into streams or rivers. The results are deplorable. Putrefaction takes place in the dumps and nuisance results, which can be detected for miles. The streams are contaminated and the fish are killed. On account of these primitive methods of waste disposal, the average sisal factory is a most depressing and disagreeable spot. Further, the water used in these operations—which has to be obtained at great expense by sinking wells or bore-holes, by making dams or reservoirs, and then raised by pumping plants—is allowed to run to waste. Two of the pressing problems, therefore, of the sisal industry are: (1) the conversion of the solid residues, including the short fibres, into some useful product like humus, and (2) the use of the waste water for raising irrigated crops.

These two problems have been successfully solved on Major Grogan's estate at Taveta in Kenya, by the Manager, Major S. C. Layzell, M.C. The work began in 1935 and has been steadily developed since. An account of the results was published in the *East African Agricultural Journal* of July 1937.

The first problem was to separate the liquid in the flume waste from the solids. At Taveta all the waste from the decorticator is passed over a grid at the end of the flume. The grid retains the solids and allows the acid 'soup' to pass into a concrete sump below, from which it is carried by a suitably graded channel, with a fall of 1 in 1,000, to the irrigated area. From the grid the solid waste is moved on slatted trucks (the usual four-wheel frame constructed of timber with a platform of slats arranged at right angles to the track) to a concrete basin where they are allowed to drain The drainage water from this basin is led by a small irrigation furrow to another area where it is utilized for growing crops. There are thus two sources of irrigation water—the main flume water and the drainage from the loaded trucks (Plate I).

The lay out of the composting ground is important. Sisal poles, in groups of four, equally spaced, are arranged on both sides and at right angles to the rail track. On these poles, placed a foot apart for providing aeration from below, the waste is lightly spread to a height of 2 feet in heaps measuring 15 feet by 4 feet. As all new heaps require a starter, any estate making compost for the first time should obtain a small supply of freshly made humus from some other sisal estate which has adopted the Indore Process. A few handfuls of this old compost, distributed evenly in the heaps, is sufficient to start fermentation. The waste is left on the poles for thirty days during which the breaking-down process, by means of thermophyllic bacteria, begins. The temperature rises to a point where it is impossible to bear one's hand in the heap.

The first turn takes place thirty days after the first formation of the heap when the contents of two heaps are run together into one, as by this time the volume has considerably decreased. After the first turn the decomposition is carried a stage further, mainly by fungi. During this phase the whole heap is often covered each morning with a long-stemmed toadstool (Plate II).

At the end of another thirty days the second turn takes place. The ripening process then begins and is completed about the ninetieth day after the original heaps were made. Major Layzell writes that the final product resembles first-class leaf mould and contains 1·44 per cent. of nitrogen. On the basis of its chemical composition alone the compost has been valued locally at £2 a ton.

A large portion of the humus is devoted to the sisal nurseries in order that all new areas can be started with vigorous and properly grown plants. The remainder finds its way to the areas producing sisal.

PLATE I

CONVERSION OF SISAL WASTE

A. Filtering the waste.   *B.* Draining.   *C.* Irrigation with waste-water

The sisal plant only does well on fertile soil and therefore needs intensive rather than extensive cultivation. Whenever this is forgotten the enterprise ends in bankruptcy for the reason that, as the soil near the factory is exhausted, the lead to the decorticators soon eats up the profit. The game is no longer worth the candle. The conversion of the wastes into humus will therefore solve this problem: the fertility of the land round the factory can be maintained and even improved. Further, the dumps of repulsive sisal waste will be a thing of the past.

The labour employed in dealing with the waste and turning the heaps from a decorticator producing 120 tons of fibre per month is thirty-four with two head men. Sixteen additional men were taken up on the grid and with the trucks, so that a labour force of fifty in all with two head men was needed for making compost at this centre. There is no difficulty in handling sisal wastes provided the workmen are given a supply of some cheap oil for protecting the skin, otherwise the juice of the leaves produces eczema on the arms and legs of those engaged on the work.

The flume liquid is mainly used for growing food crops for the labour force so as to improve the usual set ration of mealie-meal, beans, and salt. The psychological effect of all this on the labourers has been remarkable: the spectacle of a large area of bananas, sugar-cane, citrus plants, and potatoes removes all fear of a possible lack of food from the minds of the workers and their families: they feel safe. Further, their physical health and their efficiency as labourers rapidly improve. A guaranteed food-supply has proved a great attraction to labour and has provided a simple and automatic method of recruitment.

At Taveta the soil contains a good deal of lime so that the prior neutralization of the acid irrigation water is unnecessary. On other estates this point might have to receive attention. Perhaps the easiest way to get rid of the acidity would be to add sufficient powdered crude limestone to the flume water just after the solids have been separated for composting.

Two conditions must be fulfilled before the methods worked out at Taveta can be adopted elsewhere: (1) there must be a suitable area of flat land near the factory for growing irrigated crops; (2) the general layout must be such that there is ample room for a composting ground to which the wastes can be taken by a light railway and from which the finished humus can be easily transported to the irrigated area and to the rest of the estate.

One obvious improvement in the manufacture of humus on a sisal estate must be mentioned. Animal residues must be added to the vegetable wastes. If it is impossible to maintain sufficient live stock for all the sisal waste, two grades of humus should be made: first grade with animal manure for the parent plants and the nurseries, second grade for the plants which yield fibre.

MAIZE

One of the great weaknesses in British agriculture at the moment is the dependence of our live stock—such as pigs, poultry, and dairy animals—on imported foodstuffs. Our animal industry is becoming just as unbalanced as regards the supply of nutriment grown on fertile soil as our urban population. One of the animal foods imported in large quantities is maize. Unfortunately a large proportion of this import is being grown on worn-out soils. We are feeding our animals and indirectly ourselves on produce grown anywhere and anyhow so long as it is cheap.

Mother earth, however, has registered an effective protest. The maize soils of such areas as Kenya and Rhodesia soon showed signs of exhaustion. The yields fell off. Any one who has had any practical experience of maize growing could have foretold this. This crop requires a fertile soil.

The maize growers of Kenya, Rhodesia, and South Africa soon learnt this lesson. The constant cropping of virgin land with an exhausting crop rapidly reduced the yield. This happened just as the Indore Process was devised. Its application to the maize fields of Kenya and Rhodesia led to good results. The composting of the maize stalks and other vegetable residues, including green-manure crops, was taken up all over Kenya and Rhodesia.

Two examples out of many of the results which are being obtained may be quoted: at Rongai in Kenya, Mr. J. E. A. Wolryche Whitmore has adopted the Indore Process on three farms. The working oxen are being bedded down during the

PLATE II

CONVERSION OF SISAL WASTE

A. Light railway and foundation of sisal poles. *B.* Spreading. *C.* Turned
heaps with layers of elephant grass

night with dry maize stalks, wheat-straw, grass, and other roughage available. After a week under the cattle this is composted in pits with wood-ashes and earth from under the animals. If insufficient earth is used a high temperature is not maintained. Two turns at intervals of a month yield a satisfactory product after ninety days. The effect on the maize crop is very marked. In Rhodesia, Captain J. M. Moubray has obtained similar results. These are described in detail in Appendix B (p. 229).

One of the pests of maize in Rhodesia—the flowering parasite known locally as the witch-weed (*Striga lutea*)—can be controlled by humus. This interesting discovery was made by Timson whose results were published in the *Rhodesia Agricultural Journal* of October 1938. Humus made from the soiled bedding in a cattle kraal, applied at the rate of 10 tons to the acre to land severely infested with witch-weed, was followed by an excellent crop of maize practically free from this parasite. The control plot alongside was a red carpet of this pest. A second crop of maize was then grown on the same land. Again it was free from witch-weed. This parasite promises to prove a valuable censor for indicating whether or not the maize soils of Rhodesia are fertile. If witch-weed appears, the land needs humus; if it is absent, the soil contains sufficient organic matter. Good farming will therefore provide an automatic method of control.

Humus is bound to affect the quality of maize as well as the yield. In the interests both of the maize-exporting and the maize-importing countries, a new system of grading and marketing the produce of fertile soil should be introduced. Maize grown on land manured with properly made humus and without the help of artificials should be so described and graded. Only in this way can well-grown produce come into its own. It should be clearly distinguished in its journey from the field to the animal and kept separate from inferior maize. Purchasers will then know that such graded produce fed to their live stock will have been properly grown. They will soon discover that it suits their animals. This question of grading produce according to the way it is grown applies to many other crops besides maize. Its importance to the future of farming and the health of the nation is referred to in a later chapter (p. 221).

RICE

By far the most important food crop in the world is rice. It will be interesting therefore to see what response this cereal makes to humus. We should expect it would be considerable, because the rice nurseries are always heavily manured with animal manure and just before transplanting the seedlings are much richer in nitrogen than at any further stage in the life of the plant.

The first trial of the Indore Process was made by the late Mrs. Kerr at the Leper Home and Hospital, Dichpali in H.E.H. the Nizam's Dominions. Her reaction after reading *The Waste Products of Agriculture* in 1931 was: 'If he is right it will mean the utter economic revolution of India's villages.' Rice was selected as the crop on which to test the method. She died while the trial was in progress. The results are summed up in a letter from her husband, the Rev. G. M. Kerr, dated Dichpali, November 2nd, 1933, as follows:

'We have cut three and entirely average portions of our rice fields. No. 1 plot had 1·25 to 1·5 inches of Indore compost ploughed in. No. 2 plot had some farm rubbish plus ⅜ inch of Indore compost. No. 3 plot was the control and had nothing.

'Since we are eager to get these figures off to you the tabulated weight results of the straw cannot be given. Plot No. 1 was cut 12 days ago; plot No. 2 only 2 days ago, and plot No. 3 yesterday. No. 1, therefore, is dry, and Nos. 2 and 3 are still wet. We have given the straw results in similar sized bundles, but No. 1 is the better straw and will make considerably better buffalo fodder (Table 4).

'Once we get all our 30 acres of rice fields fully composted we shall be able to welcome 50 or 60 more lepers here for cleansing. This is not a scientific conclusion according to your usual methods of reckoning, but it is the practical issue as it appeals to us.'

In a subsequent letter dated October 10th, 1935, the Dichpali experience of the Indore Process was summed up as follows:

'Indore compost is one of the material blessings of this life, like steam, electricity and wireless. We simply could not do without it here. It has transformed all our agricultural interests. We have 43 acres under wet cultivation, and most of the land three years ago was of the poorest nature, large patches of it so salty that a white alum-like powder lay on the surface. We have now recovered 28 acres, and on these we are having a bumper crop of rice this year. There have never been such crops grown on the land, at least not for many years. The remaining

15 acres are as before with the rice scraggy and thin. By means of our factory of 30 pits we keep up a supply of compost, but we can never make enough to meet our needs. We are now applying it also on our fields of forage crops with remarkable results. Compost spread over a field to the depth of about one quarter of an inch ensures a crop at least three or four times heavier than otherwise could be obtained.'

TABLE 4

Crop results of three plots of rice grown under varying conditions at the Home for Lepers, Dichpali

	No. 1 Plot	No. 2 Plot	No. 3 Plot
Amount of land measured for the contrast. All portions had the same cultivation	6,364 sq. ft.	6,364 sq. ft.	6,364 sq. ft.
Amount of seed sown. All the seed sown was the same quality . . .	6 lb.	6 lb.	6 lb.
Amount of rice taken in each case by measure, not weight . . .	422 lb.	236 lb.	60 lb.
Amount of straw in similar sized bundles	138 bundles	106 bundles	40 bundles

The marked response of rice to organic matter in the rice nurseries is well known. The Dichpali results prove that the transplanted crop also responds to humus. In the nurseries the soil conditions are aerobic: after transplanting, the roots of the crops are under water, when the oxygen supply largely depends on the activities of algae. How does humus influence the rice plant in water culture under conditions when the active oxygen must be dissolved in water? Do the roots of rice in the nurseries and also after transplanting exhibit the mycorrhizal relationship? If they do, the explanation of the Dichpali results is simple. If they do not, how then does humus in wet cultivation influence photosynthesis in the green leaf? Nitrification of the organic matter would seem difficult under such conditions for two reasons: (1) the process needs abundant air; (2) the nitrate when formed would undergo excessive dilution by the large volume of water in the rice fields. If, however, the mycorrhizal association occurs in transplanted rice, the Dichpali results explain themselves.

While this book was passing through the press specimens of surface roots of transplanted rice, 116 days from the date of sowing, grown in soil manured with humus, were collected on October

27th, 1939, by Mr. Y. D. Wad in Jhabua State, Central India. They were examined by Dr. Ida Levisohn on December 11th, 1939, whose report reads as follows:

'The stouter laterals of the first order show widespread endotrophic mycorrhizal infection, the mycorrhizal regions being indicated macroscopically by opacity, beading and the absence of root hairs. The active hyphae are of wide diameter; they pass easily through the cell walls and form coils, vesicles and arbuscles; they show the early and later stages of digestion. The resulting mass of granular material appears to be rapidly translocated from the cells.'

There is no doubt that rice is a mycorrhiza-former, a fact which at once explains the remarkable response of this crop to humus and which opens up a number of new lines of investigation. Yield, quality, disease resistance, as well as the nutritive value of the grain will in all probability be found to depend on the efficiency of the mycorrhizal association.

VEGETABLES

One of the chief problems in market gardening in the open and under glass is the supply of humus. In the past, when horse transport was the rule and large numbers of these animals were kept in the cities, it was the custom, near London for example, for the wagons which brought in the crates of vegetables for the early morning market to take back a load of manure. The introduction of the internal combustion engine changed this: a general shortage of manure resulted. In most cases market gardens are not run in connexion with large mixed farms, so there is no possibility of making these areas self-supporting as regards manure: the essential animals do not exist. The result is that an increasing proportion of the vegetables sold in the cities is raised on artificial manure. In this way a satisfactory yield is possible, but in taste, quality, and keeping properties the product is markedly inferior to the vegetables raised on farm-yard manure.

It is an easy matter to distinguish vegetables raised on NPK. They are tough, leathery, and fibrous: they also lack taste. In marked contrast those grown with humus are tender, brittle, and possess abundant flavour. One of the lessons in dietetics which should be taught to children in every school and institution in the country, and also in every home, should be the difference between vegetables, salads, potatoes, and fruit grown with humus or with artificials. Evidence is accumulating that liability to common ailments like colds, measles, and so forth becomes much less when the vegetables, fruit, potatoes, and other food consumed are raised from fertile soil and eaten fresh.

How is the necessary humus for the high-quality vegetables needed by urban areas to be obtained? Two solutions of the problem are possible.

In the first place, market gardening should, whenever possible, be conducted as a branch of mixed farming with an adequate head of live stock, so that all the waste products, vegetable and animal, of the entire holding can be converted into humus by the Indore Process. The first trial of this system was carried out at the Iceni Nurseries, Surfleet, Lincolnshire. The work commenced in December 1935 and can best be described in Captain Wilson's own words taken from a memorandum he drew up for the members of the British Association who visited his farm on September 4th, 1937:

'The Iceni Estate consists of about 325 acres comprised as follows:

Arable land, &c.	225 acres
Permanent grassland	30 ,,
Rough wash grazings	35 ,,
Land under intensive horticulture	35 ,,

'The main idea in the development of the estate has been to prove that even to-day, in certain selected areas of England, it is a commercial proposition to take over land which has been badly farmed, and bring it back to a high state of fertility, employing a large number of persons per acre.

'To this end the estate has been developed as a complete agricultural unit with a proper proportion of live stock, arable land, grass land and horticulture, with the belief that after a few years of proper management the estate can become very nearly, if not entirely, a self-supporting unit, independent of outside supplies of chemical manures, &c., and feeding stuffs, the land being kept in a high state of fertility, which is quite unusual to-day, by:

(1) A proper balance of cropping.
(2) The conversion of all wheat straw into manure in the crew yards and the utilization of this manure and as much as possible of the waste products of the land for making humus for the soil.

'As regards (2), the method of humus-making which has been employed is known as "The Indore Process", and it has proved remarkably successful. The output in 1936 amounted to approximately 700 tons, and in the current year will probably be about 1,000 tons.

'As a result of this utilization of humus, the land under intensive cultivation has already reached a state of independence, and for the last two years *no chemicals have been used in the gardens at all either as fertilizers or as sprays for disease and pest control.* The only wash which has been used on the fruit trees is one application each winter of lime sulphur, and it is hoped to eliminate this before long.

'The farm land is not yet independent of the purchase of fertilizers, but the amount used has been steadily reduced from 106 tons used in 1932, costing £675, to 40½ tons in the current year, costing £281. Similarly the potato crop, which formerly was sprayed four or five times, is now only sprayed once, and this it is hoped will also be dispensed with before many years when the land has become healthy and in a proper state of fertility.

'Eventually, with a properly balanced crop rotation, there is no doubt in my mind that the same degree of independence can be reached on the farm as has already been attained on my market-garden land.

'The probable cropping will eventually work out as follows:

75 acres potatoes.
75 acres wheat.
25 acres barley, oats, beans and linseed (for stock feeding).
15 acres roots (for stock feeding).
30 acres one year clover and rye-grass leys for feeding with pigs and
   poultry and cutting for hay, ploughing in the aftermath.

'The live stock carried on the farm at the June returns was as follows:

22 cattle (cows and young stock of my own breeding).
14 horses (including foals).
15 sows (for breeding).
103 other pigs.
120 laying hens (of my own stock).

'And although it is rather early to say, I believe that the above figures may be about right for the size of the farm, with the addition of about 20 cattle for winter yard feeding. This latter importation will be rendered unnecessary in a few years when the number of cattle of my own breeding will have increased.'

Since this memorandum a further advance has been made at Surfleet. The factor which at present limits production on the alluvial soils in the Holland Division of Lincolnshire (in which Captain Wilson's vegetable garden is situated) is undoubtedly soil aeration. These soils pack easily, so the supply of oxygen for the micro-organisms in the soil and for the roots of the crops is frequently interrupted. Sub-soil drainage tends to reduce this adverse factor. During the autumn of 1937 the whole of Captain Wilson's vegetable area was pipe drained. As was expected, the improvement in soil aeration which instantly followed has enabled the crops to obtain the full benefit of humus. Here is a definite example where the establishment of Nature's equilibrium between the soil, the plant, and the animal has resulted in increased crops and in higher quality produce.

Another and perhaps a simpler solution of the organic matter problem in vegetable growing is to make use of the millions of tons of humus in the controlled tips in the neighbourhood of our cities and towns. This subject is discussed in detail in Chapter VIII.

VINE

A comparison between the cultivation of the vine in the East and the West is interesting in more than one respect. In the Orient this crop is grown mostly for food: in the Occident, including Africa, most of the grapes are made into wine.

The feature of the cultivation of the vine in Asia is the long life of the variety, the universal use of animal manure, and the comparative absence of insect and fungous diseases. Artificial manures, spraying machines, and poison sprays are unknown.

In the West the balance between the area under vines and the number of live stock has been lost: the vine has largely displaced the animal: the shortage of farm-yard manure has been made up by the chemical fertilizer: the life of the variety is short: insect and fungous diseases are universal: the spraying machine and the poison spray are to be seen everywhere: the loss of balance in grape growing has been accompanied by a lowering in the quality of the wine.

During the last three summers, in the course of extensive tours in Provence, a sharp look-out was kept for vineyards in which the appearance of the vines tallied in all respects with those of Central Asia, namely, well-grown plants which looked thoroughly at home, and in which the foliage and young growth possessed real bloom. At last near the village of Jouques in the Department of Bouches du Rhône such vines were found. They had never received any artificials, only animal manure: the vineyard had a good local reputation for the quality of its wine. Arrangements were made with the proprietress to have the active roots examined. They exhibited the mycorrhizal association. The vine is a mycorrhiza-former and therefore humus in the soil is essential for perfect nutrition; the long life of the variety and the absence of disease in Central Asia are at once explained.

In a recent survey of fruit growing in the Western Province of the Union of South Africa, which appeared in the *Farmer's Weekly* (Bloemfontein) of August 23rd, 1939, Nicholson refers to a local vineyard, on the main road between Somerset West and Stellen-bosch, which has taken up the Indore Process:

'Motorists travelling along this road cannot help noticing how healthy this farmer's vineyards look and how orderly is the whole farm. Early this winter I visited it in time to see the huge stacks of manure—beautiful, finely rotted bush which had been helped to reach that state by being placed in the kraal under the animals. Pigs had played their part too. During the wine-pressing season all the skins of the grapes are fed to the pigs and later returned to the vineyards in the form of manure.'

When the vine growers of Europe realize how much they are losing by an unbalanced agriculture in the shape of the running out of the variety, loss of resistance to disease, and loss of quality in the wine, steps will no doubt be taken by a few of the pioneers to increase the head of live stock, to convert all the available wastes into humus, and to get back to Nature as quickly as possible.

BIBLIOGRAPHY

DYMOND, G. C. 'Humus in Sugar-cane Agriculture', *South African Sugar Technologists*, 1938.

HOWARD, A. 'The Manufacture of Humus by the Indore Process', *Journal of the Royal Society of Arts*, lxxxiv, 1935, p. 25 and lxxxv, 1936, p. 144.

—— 'Die Erzeugung von Humus nach der Indore-Methode', *Der Tropenpflanzer*, xxxix, 1936, p. 46.

—— 'The Manufacture of Humus by the Indore Process', *Journal of the Ministry of Agriculture*, xlv, 1938, p. 431.

—— 'En Busca del Humus', *Revista del Instituto de Defensa del Café de Costa Rica*, vii, 1939, p. 427.

LAYZELL, S. C. 'The Composting of Sisal Wastes', *East African Agricultural Journal*, iii, 1937, p. 26.

TAMBE, G. C., and WAD, Y. D. 'Humus-manufacture from Cane-trash', *International Sugar Journal*, 1935, p. 260.

CHAPTER VI

DEVELOPMENTS OF THE INDORE PROCESS
GREEN-MANURING

SINCE Schultz-Lupitz first showed about 1880 how the open sandy soils of North Germany could be improved in texture and in fertility by the incorporation of a green crop of lupins, the possibilities of this method of enriching the land have been thoroughly explored by the Experiment Stations. After the role of the nodules on the roots of leguminous plants in the fixation of atmospheric nitrogen was proved, the problems of green-manuring naturally centred round the utilization of the leguminous crop in adding to the store of combined nitrogen and organic matter in the soil. At the end of the nineteenth century it seemed so easy, by merely turning in a leguminous crop, to settle at one stroke and in a very economical fashion the great problem of maintaining soil fertility. At the expenditure of a little trouble, the leguminous nodule might be used as a nitrogen factory while the remainder of the crop could provide humus. All this might be accomplished at small expense and without any serious interference with ordinary cropping. These expectations, a natural legacy of the NPK mentality, have led to innumerable green-manuring experiments all over the world with practically every species of leguminous crop. In a few cases, particularly in open, well-aerated soils where the rainfall after the ploughing in of the green crop was well distributed and ample time was given for decay, the results have been satisfactory. In the majority of cases, however, they have been disappointing. It will be useful, therefore, to examine the whole subject and to determine if possible the reasons why this method of improving the fertility of the soil seems so often to have failed.

A consideration of the factors involved in the growth, decay, and utilization of the residues of a green crop will at once explain the general failure of green-manuring to increase the following crop and also put an end for all time to the somewhat extravagant hopes of repeating the German results, which succeeded because all the factors, including time, happened to be favourable. It is no use slavishly copying this method unless we can at the same time reproduce the North German soil and climatic conditions.

The chief factors in green-manuring are: (1) a knowledge of the nitrogen cycle in relation to the local agriculture; (2) the conditions necessary for rapid growth and also for the formation of abundant nodules on the roots of the leguminous crop used for green-manuring; (3) the chemical composition of the green crop at the moment it is ploughed in; (4) the soil conditions during the period when decay takes place. These four factors must be studied before the possibilities of green-manuring can be explored.

The importance of the nitrogen cycle in relation to the local agriculture is a factor in green-manuring to which far too little attention has been paid. As will be shown more fully in Chapter XIV, the full possibilities of green-manuring can only be utilized when we know at what periods of the year nitrate accumulations take place, how these accumulations fit in with the local agricultural practice, and when nitrates are liable to be lost by leaching and other means. If the crop does not make the fullest use of nitrate, this precious substance must be immobilized by means of green-manure or by means of weeds and algae. It must not be left to take care of itself. It must either be taken up by the crop or banked by some other plant.

The soil conditions necessary for the growth of the leguminous crop used as a green-manure have never been sufficiently studied. Clarke found at Shahjahanpur in India that it was advantageous to apply a small dressing of farm-yard manure to the land just before the green crop is sown. The effect of this is to stimulate growth and nodular development in a remarkable way. Further, the green crop when turned in decays much faster than when this preliminary manuring is omitted. It may be that besides stimulating nodular development the small dressing of farm-yard manure is necessary to bring into effective action the mycorrhizal association which is known to exist in the roots of most leguminous plants. This association is a factor which has been completely forgotten in green-manuring. There is no reference to it in Waksman's excellent summary on pp. 208–14 of the last edition of his monograph on humus. This factor will probably also prove to be important in the utilization of the humus left by green-manuring.

The living bridge between the humus in the soil and the plant must be properly fed, otherwise the nutrition of the crop we wish to benefit is almost certain to suffer.

As growth proceeds the chemistry of a green crop alters very considerably: the material in a young or in a mature crop, when presented to the micro-organisms of the soil, leads to very different results. Waksman and Tenney have set out the results of the decomposition of a typical green-manure plant (rye) harvested at different periods of growth. When the plants are young they decompose rapidly: a large part of the nitrogen is released as ammonia and becomes available. When the plants are mature they decompose much more slowly: there is insufficient nitrogen for decay, so the micro-organisms utilize some of the soil nitrates to make up the deficiency. Instead of enriching the soil in available nitrogen the decay of the crop leads to temporary impoverishment. These fundamental matters are summed up in the following Table:

TABLE 5

Rapidity of decomposition of rye plants at different stages of growth
(Waksman and Tenney)

Two grammes of dry material decomposed for 27 days

Stage of growth	CO_2 given off	Nitrogen liberated as ammonia	Nitrogen consumed from the medium
	mg. C	mg. N	mg. N
Plants only 25–35 cm. high . . .	286·8	22·2	0
Just before heads begin to form . . .	280·4	3·0	0
Just before bloom .	199·5	0	7·5
Plants nearly mature .	187·9	0	8·9

The amount of humus which results from the decay of a green crop also depends on the age of the plants. Young plants, which are low in lignin and in cellulose, leave a very small residue of humus. Mature plants, on the other hand, are high in cellulose and lignin and yield a large amount of humus. These differences are brought out in Table 6 on p. 90.

It follows from these results that if we wish to employ green-manuring to increase the soil nutrients quickly, we must always plough in the green crop in the young stage; if our aim is to increase the humus content of the soil we must wait till the green-manure crop has reached its maximum growth.

The soil conditions after the green crop is ploughed in are no less important than the chemical composition of the crop. The

TABLE 6

Formation of humus during decomposition of rye plants at different stages of development (Waksman and Tenney)

Chemical constituents	At beginning of decomposition*	At the end of decomposition period†	
JUST BEFORE HEADS BEGIN TO FORM			
	mg.	mg.	% original
Total water-insoluble organic matter .	7,465	2,015	27·0
Pentosans . .	2,050	380	18·5
Cellulose. . .	2,610	610	23·4
Lignin . . .	1,180	750	63·6
Protein insoluble in water . . .	816	253	31·0
PLANTS NEARLY MATURE			
	mg.	mg.	% original
Total water-insoluble organic matter .	15,114	8,770	58·0
Pentosans . .	3,928	1,553	39·5
Cellulose. . .	6,262	2,766	44·2
Lignin . . .	3,403	3,019	88·7
Protein insoluble in water . . .	181	519	286·7

* 10 gm. material (on dry basis) used for young plants and 20 gm. for old plants.
† 30 days for young plants and 60 days for mature plants.

micro-organisms which decay the green-manure require four things: (1) sufficient combined nitrogen and minerals; (2) moisture; (3) air; (4) a suitable temperature. These must all be provided together.

The factor which so often leads to trouble is the poverty of the soil—insufficient combined nitrogen and minerals. It follows, therefore, that when a mature crop is ploughed in the effect of its decay on the next crop will always depend on the fertility of the soil. If the soil is in a poor condition most of the combined nitrogen available will be immobilized for the decay of the green-manure;

the next crop will suffer from starvation; green-manuring will then be a temporary failure. If, however, the soil is fertile or if we plough in freshly prepared humus with the green crop, the extra combined nitrogen needed for decay will then be present; the next crop will not suffer. Soil fertility in this, as in so many other matters, gives the farmer considerable latitude. All sorts of things can be done with perfect safety with a soil in good heart which are out of the question when the soil is infertile. A good reserve of fertility, therefore, will always be an important factor in green-manuring.

As the decomposition of a green crop is carried out by micro-organisms, decay ceases if the moisture falls below a certain point.

Again, if the air supply is cut off by excessive rain after plough-ing in or by burying the green crop too deeply, an anaerobic soil flora rapidly develops which proceeds to obtain its oxygen supply from the substratum. The valuable proteins are attacked and their nitrogen is released as gas. The chemical reactions of the peat bog replace those of the early stages of a properly managed compost heap. This frequently happens under monsoon conditions and is one of the reasons why green-manuring is so often unsatisfactory in tropical agriculture.

Finally, the temperature factor is important in countries like Great Britain which have a winter. Here green-manure crops must often be turned in during the autumn before the soil gets too cold, so that the early stages of decay can be completed before winter comes.

The uses of green-manuring in agriculture can now be con-sidered. Generally speaking they fall into three classes: (1) the safeguarding of nitrate accumulations; (2) the production of humus, and (3) a combination of both.

THE SAFEGUARDING OF NITRATE ACCUMULATIONS

In studying this important matter we must at the outset consider how Nature, if left to herself, always deals with the nitrates pre-pared from organic matter by the micro-organisms in the soil. They are never allowed to run to waste but are immobilized by plants including the film of algae in the surface soil. These latter are easily decomposed: they are therefore exceedingly valuable agencies for safeguarding nitrates.

The farmer has at his command two methods of nitrate immobilization. He can either intercept his surplus nitrate accumulations by sowing a leguminous crop or by managing his weeds and soil algae so that they do the same thing automatically. In either case nitrates which would otherwise run to waste are converted into young fresh growth which cannot then be lost by leaching and which later on can be rapidly converted back into available nitrogen and minerals by the organisms in the soil. Obviously if weeds can be managed so that all nitrate accumulations can be utilized and the resulting growth can be turned under and decomposed in time for the next crop, there is no need to sow a leguminous crop to do what Nature herself can do so much better.

One of the best examples I have seen of the combined use of weeds and catch crops for immobilizing nitrates was worked out by Mr. L. P. Haynes on the large hop garden of Messrs. Arthur Guinness, Son & Co. at Bodiam in Sussex. Surface cultivation in this garden ceases in August soon after the hops form. A little mustard is then sown which, with the chickweed, soon produces a green carpet without interfering with the ripening of the hops. At picking time the mixed seedlings are well established, after which they have the nitrates formed at the end of the summer and in early autumn entirely to themselves. Growth is very rapid. During the autumn sheep are brought in to graze the mustard. Their urine and dung fall on the chick-weed and so contribute a portion of the essential animal wastes. In the spring the easily decomposed chickweed is ploughed into the fertile soil and decayed in good time for the next crop of hops. The soil of this hop garden is now heavily charged with chickweed seeds so that the moment surface cultivation is stopped the following August a new crop starts. This management of a common weed of fertile soil to fit in with the needs of the hop appeared to me to be nothing short of a stroke of genius. It would be difficult to find a more efficient green-manure crop than the one Nature has provided for nothing. Could there be a better example of the use of a fertility reserve for rapidly decomposing a green crop in the early spring? The ground at Bodiam is hardly ever uncovered; it is occupied either by hops or by chickweed; one crop dovetails into the other; the energy of sunlight is almost completely utilized throughout the year; the invisible labour force of the hop garden—earthworms and micro-organisms—is kept fully occupied. As the use of artificials and poison sprays is reduced, there will be a corresponding increase in efficiency in this section of the unseen establishment.

Much more use might be made of this method of green-manuring in countries like Great Britain. In fruit, vegetable, and potato growing particularly, there seems no reason why an autumn crop of weeds should not be treated as green-manure on Bodiam lines. If the land is in good heart, the soil will have no difficulty in decaying the weeds. If the land is poor in organic matter, a dressing of freshly prepared humus of not less than 5 tons to the acre should be spread on the weeds before they are turned under.

THE PRODUCTION OF HUMUS

The production of humus, by means of a green-manure crop, is a much more difficult matter than the use of this method for immobilizing nitrates. Nevertheless, it is of supreme importance in the maintenance of soil fertility. The factors involved in the transformation of green-manure into humus in the soil are the same as those in the compost heap. All factors must operate together. Failure of one will upset the process entirely. If this occurs the next crop will be sown in soil which has been placed in an impossible condition. The land will be called upon to complete the formation of humus and to grow a crop at the same time. This is asking too much. The soil will take up its interrupted task and proceed with the manufacture of humus. It will neglect the crop. The uncontrollable factor is the rainfall. It must be just right if humus manufacture in the soil is to succeed. In India, for example, during an experience of twenty-six years it used to be just right about once in six or seven years. It was completely wrong in the remainder. Often there was too much rain after ploughing in, when the aerobic phase never developed and bog conditions were established instead. At other times there was insufficient rain for the early fungous stage. Where, however, irrigation is available, any shortage of the Indian monsoon makes no difference.

In exceptional cases, however, it is possible to carry on the manufacture of humus in the soil without any risk of temporary failure. One British example may be quoted. On some of the large farms in the Holland Division of Lincolnshire peas are grown as a rotation crop with potatoes. The problem is to manufacture humus before the next crop of potatoes is planted. This has been solved. Early in July the peas are cut and carried to the shelling machines where the green seeds are separated and large quantities of crushed haulm are left. Immediately after the removal of the peas the land is sown with beans. The crushed pea haulm is then scattered on the surface of the newly sown land followed by a light dressing of farm-yard manure—about 6 or 7 tons to the acre. The beans grow through the fermenting layer on the surface of the soil and help to keep it moist. While the beans are growing humus is being manufactured in a thin sheet all over the field. At the end of September, when the beans are in flower, this sheet composting on the ground is complete. The green crop is then lightly ploughed in together with a layer of freshly prepared compost. Humus manufacture is then continued in the soil. The beans under these conditions decay quickly; the process of humus manufacture is completed before the planting of the next potato crop.

THE SAFEGUARDING OF NITRATES FOLLOWED BY THE MANUFACTURE OF HUMUS

The immobilization of nitrates by means of a green crop followed by the conversion of the green-manure into humus needs time and complete control of all the operations. An example of the successful use of this method is described in Chapter XIV (p. 211). Heavy crops of sugar-cane were produced at Shahjahanpur in the United Provinces by intercepting the nitrates accumulated at the break of the south-west monsoon by means of a leguminous crop and then converting this into humus with the assistance of the autumn accumulation of nitrate in the same soil.

It follows from the principles underlying green-manuring and the applications of these principles to agricultural practice that the ploughing in of a green crop is not a simple question of the addition of so many pounds of nitrogen to the acre but a vast and many-sided biological problem. Moreover it is dynamic, not static; the agents involved are alive; their activities must fit in with one another, with agricultural practice on the one hand and with the season on the other. If we attempt to solve such a complex on the basis of mere nitrogen content or on that oı carbon: nitrogen ratios, we are certain to run counter to great biological principles and come into conflict with one rule in Nature after another. It is little wonder, therefore, that green-manuring has led to so much misunderstanding and to so much disappointment.

THE REFORM OF GREEN-MANURING

The uncertainties of humus manufacture in the soil can be overcome by growing the green crop to provide material for composting. This of course adds to the labour and the expense, but in many countries it is proving a commercial proposition. In Rhodesia, for example, crops of *san* hemp are now regularly grown to provide litter, rich in nitrogen, for mixing with maize stalks so as to improve the carbon: nitrogen ratio of the bedding used in the cattle kraals. In this way the burden on the soil is greatly reduced; it is only called upon to decay what is left of the root system of the green crop at harvest time. Humus manufacture is shared between the soil and the compost heap.

In converting materials low in nitrogen (such as sugar-cane leaves and cotton stalks) into humus it is an immense advantage to mix these refractory materials with some leguminous plant in the green state. The manufacture of humus is speeded up and simplified; the amount of water needed is reduced; the land on which the green crop was raised benefits.

BIBLIOGRAPHY

CLARKE, G. 'Some Aspects of Soil Improvement in relation to Crop Production', *Proc. of the Seventeenth Indian Science Congress*, Asiatic Society of Bengal, Calcutta, 1930, p. 28.
WAKSMAN, S. A., and TENNEY, F. G. 'Composition of Natural Organic Materials and their Decomposition in the Soil, *Soil Science*, xxiv, 1927, p. 275; xxviii, 1929, p. 55; and xxx, 1930, p. 143.

CHAPTER VII

DEVELOPMENTS OF THE INDORE PROCESS

GRASS-LAND MANAGEMENT

Two very different methods of approach to the problems of grass-land management in a country like Great Britain are possible. We can either study the question from the point of view of the present organization of agricultural research in this country or we can bring the world-wide experience of the grass and clover families to bear as if no institutions like the Welsh Plant Breeding Station, the Rowett Institute at Aberdeen, or the Rothamsted Experiment Station—all of which deal independently with some fragment of the grass-land problem—had ever been contemplated. As the advantages of the fresh eye are many and obvious and as the writer has had a long and extensive first-hand experience of the cultivation of a number of crops belonging to the grass and clover families, the principles underlying grass-land management in Great Britain will be considered from a new angle, namely, the conditions which practical experience in the tropics has shown to be necessary for grasses and legumes to express themselves and to tell their own story.

The grass and clover families are widely distributed and cultivated all over the world—from the tropics to the temperate zones and at all elevations and under every possible set of soil and moisture conditions—either as separate crops or more often mixed together. Everywhere the equivalent of the short ley, composed of grasses and legumes, is to be found. The successful mixed cultivation of these two groups of plants has been in operation for many centuries: in the Orient they were grown together in suitable combinations long before England emerged from the primitive condition in which the Roman invaders found it—an island covered for the most part with dense forests and impassable bogs.

What are the essential requirements of the grass and clover families? The clearest answer to this question is supplied by tropical agriculture; here the growth factors impress themselves on the plant much more definitely and dramatically than they do in a damp temperate island like Great Britain where all such reactions are apt to be very much toned down and even blurred.

Sugar-cane, maize, millets, and the *dub* grass of India (*Cynodon dactylon* Pers.) are perhaps the most widely cultivated and the most suitable grasses for this study. Lucerne, *san* hemp (*Crotalaria juncea* L.), the cluster bean (*Cyamopsis psoralioides* D.C.), and the pigeon pea (*Cajanus indicus* Spreng.) are corresponding examples of the clover family. The last two of these are almost always grown mixed either with millets or maize, very much in the same way as red clover and rye-grass are sown together in Great Britain.

The grass family must first be considered. A detailed account of the cultivation of the sugar-cane will be found in a later chapter (p. 200). Humus and ample soil aeration, combined with new varieties which suit the improved soil conditions, enable this grass to thrive, to resist disease, and to produce maximum yields and high quality juice without any impoverishment of the soil. Maize behaves in the same way and is perhaps one of our best soil analysts. Any one who attempts to grow this crop without organic matter will begin to understand how vital soil fertility is for the grass family. The requirements of the *dub* grass in India, one of the most important fodder plants of the tropics, are frequent cultivation and abundance of humus. The response of this species to a combination of humus and soil aeration is even more remarkable than in maize: once these factors are in defect growth stops. The behaviour of *dub* grass, as will be seen later on, indicates clearly what all grasses the world over need.

Any one who grows lucerne in India under irrigation will court certain failure unless steps are taken to keep the crop constantly supplied with farm-yard manure and the aeration of the surface soil at a high level. When suitable soil conditions are maintained it is possible to harvest twenty or more good crops a year. Once the surface soil is allowed to pack and regular manuring is stopped, a very different result is observed. The number of cuts falls off to three or four a year and the stand rapidly deteriorates. When *san* hemp is grown for green-manuring or for seed in India satisfactory results are only obtained if the crop is manured with cattle manure or humus. These two leguminous crops do not stand alone. Every member of this group I have grown responds at once to farm-

H

yard manure or humus. But all this is not in accordance with theory.

According to the text-books the nodules in the roots of leguminous plants should be relied on to furnish combined nitrogen and this group should not need nitrogenous manure. Practical experience and theory are so wide apart as to suggest that some other factor must be in operation. It was not till January 1938 that I discovered what this factor was. On the Waldemar tea estate in Ceylon I saw a remarkable crop of a green-manure plant—*Crotalaria anagyroides*—growing in soil rich in humus. The root development was exceptional: an examination of the active roots showed that they were heavily infected with mycorrhiza. Other tropical leguminous plants growing in similar soils also exhibited the mycorrhizal association. So did several species of clover collected in France and Great Britain. These results at once suggested the reason why *san* hemp, lucerne, and many other tropical legumes respond so strikingly to cattle manure. They must all be mycorrhiza-formers.

The fact that leguminous plants and grasses respond to the same factors and that the former group are mycorrhiza-formers suggested that this association would also be found in the grasses. Sugar-cane was first investigated. It proved to be a mycorrhiza-former. The grasses of the meadows and pastures of France and Great Britain were then studied. The herbage of the celebrated meadows of La Crau, between Salon and Arles in Provence, was examined for mycorrhiza in 1938 and again in 1939. In both seasons the roots of the grasses were found to be infected with mycorrhiza. Dr. Levisohn's report on the samples collected in July 1939 reads as follows: 'Sporadic but deep infection of the long and short roots: coarse mycelium inter- and intra-cellular: digestion stages: the products of digestion seem to be translocated rapidly.' In the material from La Crau examined in 1939 the most remarkable example of the mycorrhizal relationship occurred in a species of *Taraxacum* which formed at least a quarter of the herbage. Here the infection of the inner layers of the long and short roots was 'very widespread and deep. The mycelium is of large diameter, thin-walled with granular contents. Distribution mainly intra-cellular. Digestion showing all stages of disintegration. Root hairs sparsely formed. The mycorrhizal regions of the roots are indicated macroscopically by beading, greater opacity, and slight yellowing of the infected zones' (Levisohn). This suggests that some or all of the so-called weeds of grass-land may well play an important role in the transmission of quality from soil to plant and in the nutrition of the animal. Samples of the turf from two well-known farms in England—Mr. Hosier's land in Wiltshire and Mr. William Kilvert's pastures in Corve Dale in Shropshire— were then examined. They gave similar results to those of La Crau. Clearly the grass family, like the clover group, are mycorrhiza-formers, a fact which at once explains why both these classes of plants respond so markedly to humus.

This independent approach to the grass-land problems of countries like Great Britain has brought out new principles. Grasses and clovers fall into one group as regards nutrition and not, as hitherto thought, into two groups. Both require the same things—humus and soil aeration. Both are connected with the organic matter in the soil by a living fungous bridge which provides the key to their correct nutrition and therefore to the management of grass-land. If this view is a sound one it follows that any agency which will increase the natural formation of humus under the turf of our grass-lands will be followed by an improvement in the herbage and by an increase in their stock-carrying capacity. The methods which increase humus formation in the soil must now be considered. The following may be mentioned:

1. *The bail system.* The most spectacular example of humus manufacture in the soil underneath a pasture is that to be seen on Mr. Hosier's land on the downs near Marlborough. By a stroke of the pen, as it were, he abolished the farm-yard, the cowshed, and the dung-cart in order to counter the fall in prices which followed the Great War. He reacted to adversity in the correct manner: he found it a valuable stimulant in breaking new ground. The cows were fed and made to live out of doors. They were milked in movable bails. Their urine and dung were systematically distributed at little cost over these derelict pastures. The vegetable residues of the herbage came in contact with urine, dung, air, water, and bases. The stage was set for the Indore Process. Mr. Hosier's invisible labour force came into action: the micro-

organisms in the soil manufactured a sheet of humus all over the downs: the earthworms distributed it. The roots of the grasses and clovers were soon geared up with this humus by means of the mycorrhizal association. The herbage improved; the stock-carrying capacity of the fields went up by leaps and bounds. Soil fertility accumulated; every five years or so it was cashed in by two or three straw crops; another period under grass followed, and so on. Incidentally the health of the animals also benefited; the prognostications of the neighbourhood (when this audacious innovation started) that the cows and heifers would soon perish through tuberculosis and other diseases have not been fulfilled.[1]

2. *The use of basic slag.* On many of the heavy soils under grass the limiting factor in humus production is not urine but oxygen. Everything except air is there in abundance for making humus— vegetable and animal wastes as well as moisture. Under such turf the land always suffers from asphyxiation. The soil dies. This is indicated by the absence of nitrates under such turf. About fifty years ago it was discovered that such pastures could be improved by dressings of basic slag. As this material contains phosphate, and as its use stimulates the clovers, it was assumed that these soils suffered from phosphatic depletion as a result of feeding a constant succession of live stock, each generation of which removes so many pounds of phosphate in their bones. When, however, we examine the turf of a slagged pasture we find that humus formation has taken place. If the application of slag is repeated on these heavy lands after an interval of five or six years there is often no further response. When we apply basic slag to pastures on the chalk there is no result. There is phosphate depletion on strong lands only at one point; none at all on light chalk downs. These results do not hold together; indeed they contradict one another. Are we really dealing with phosphate deficiency in these lands? May not the humus formed after slag is added explain the permanent benefit of this manuring? May it not prove that the effect of slag on heavy soils has been in the first instance a physical one which has improved the aeration, reduced the acidity, and so

[1] Mr. Hosier has done more than solve a local problem and provide evidence in support of a new theory. His work has drawn attention to the potential value of our downlands—areas which in Roman and Saxon times supported a large proportion of the population of Great Britain.

helped humus manufacture to start? We can begin to answer these questions by studying what happens when the aeration of heavy grass-land is improved by an alternative method—sub-soiling.

3. *Sub-soiling*. The effect of sub-soiling heavy grass-lands was described by Sir Bernard Greenwell, Bt. in a paper read to the Farmers' Club on January 30th, 1939, in the following words:

'Taking our grass-land first, probably more can be done by proper mechanical treatment followed by intensive stocking than by artificial manuring. Some people are suggesting that we should plough up a lot of our second-rate pasture land and re-sow it, but this I have found is very speculative as the cost is in the neighbourhood of £3 to £5 an acre and the results are bound to be uncertain. By cleaning out ditches, reopening drains and by mole draining, however, a lot can be done. I have also found that by using a Ransome mole plough or sub-soiler of the wheel type, pulled through the land at a depth of 12 inches to 14 inches, 4 feet apart, one can produce much better grass, and this is proved by the greatest expert of all—the animal. In a field which was partly sub-soiled we found that this sub-soiled part was grazed hard by the cattle, and the part that was not treated in this way was only lightly picked over. The cost of this is about 2s. 6d. per acre without overheads and lost time. We reckon £1 a day for a 40-h.p. tractor, including labour, depreciation, &c., and a tractor will do 9 to 10 acres a day sub-soiling at 4 feet intervals.'

Poor aeration was obviously the limiting factor at Marden Park. Once this was removed humus formation started and the herbage improved. It will be interesting to watch the results of the next stage of this work. Half of a sub-soiled field has been dressed with basic slag and the reaction of the animals is being watched. If they graze the field equally, basic slag is probably having no effect: if the animals prefer the slagged half then this manure is required.[1]

4. *The cultivation of grass-land*. One of the recommendations of the Welsh Plant Breeding Station is the partial or complete cultivation of grass-land. Partial cultivation is done from the surface by various types of harrow: complete cultivation by the plough. In

[1] The Marden Park results suggest a further question. Will sub-soiling at 2s. 6d. an acre replace the ploughing-up campaign recently launched by the Ministry of Agriculture for which the State pays £2 an acre? If, as seems likely, the basic slag and ploughing-up subsidies are both unnecessary, a large sum of money will be available for increasing the humus of the soils of Great Britain, the need for which requires no argument.

both cases aeration is improved; the production of humus is stimulated; generally speaking the result obtained is in direct proportion to the degree of cultivation; ploughing up and re-seeding is far better for the grass than mere scarification with harrows. In this work we must carefully distinguish the means and the end. The agency is some form of cultivation; the consequence is always the manufacture of humus.

It will be evident that the various methods by which humus is manufactured under the growing turf itself or by ploughing up and rotting the old turf agree in all respects with what is to be learned from the grasses and the legumes of the tropics. Sir George Stapledon's advice as regards Great Britain is supported by the age-long experience of the agriculture of the East. No stronger backing than this is possible. There is only one grass-land problem in the world. It is a simple one. The soil must be brought back to active life. The micro-organisms and earthworms must be supplied with freshly made humus and with air. Varieties of grasses and legumes which respond to improved soil conditions must then be provided. In this way only can the farmers of Great Britain make the most of our green carpet. Our grass-lands will then be able to do what Nature does in the forest—manure themselves.

The order in which improvements should be introduced in grass-land management is important. *Soil fertility must first be increased* so that the grasses and clovers can fully express themselves. Improved varieties should then be selected to suit the new soil conditions. If we study the variety by itself without any reference to the soil and develop higher yielding strains of grasses and clovers for the land as it is now, there is a danger, indeed almost a certainty, that the farmer will be furnished with yet another means of exhausting his soil. The new varieties will have a short life: they will prove to be a boomerang: the last state of the farm will be worse than the first. If, however, the soil conditions are first improved and the system of farming is such that soil fertility is maintained, the plant breeder will be provided with a safe field for his activities. His work will then have a permanent value.

How are we to test the fertility of grass-land? Mr. Hosier has supplied the answer. Grass-land can be tested for fertility by means of a complete artificial manure. If the soil is really fertile, such a dressing will give no result, because no limiting factor in the shape of shortage of nitrogen, phosphorus, or potash exists. Mr. Hosier has summed up his experience of this matter in a letter dated Marlborough, April 6th, 1938, as follows:

'On my improved grass-land, I have on several occasions put down experimental plots of artificial manures and there was no response even where there was a complete fertilizer applied. Before I started open-air dairying on a big scale in 1924, I put down 150 plots and in many places I could write my name with artificials.'

The value of this experience does not end with the testing of soil fertility. It indicates the very high proportion of the grass-lands of western Europe which are infertile and which need large volumes of humus to restore their fertility. Most of the fields under grass will respond to artificials. All these are infertile.

The consequences of the improvement of grass-land in a country like Great Britain can now be summarized. The land will carry more live stock. The surplus summer grass can be dried for winter feeding. The stored fertility in the pastures can be cashed in at any time in the form of wheat or other cereals. A valuable food reserve in time of war will always be available. As Mr. Hosier has shown, there will be no damage from wireworms when such fertile pastures are broken up and sown with wheat.

BIBLIOGRAPHY

GREENWELL, SIR BERNARD. 'Soil Fertility: The Farm's Capital', *Journal of the Farmers' Club*, 1939, p. 1.

HOSIER, A. J. 'Open-air Dairying', *Journal of the Farmers' Club*, 1927, p. 103.

HOWARD, A. *Crop Production in India: A Critical Survey of its Problems*, Oxford University Press, 1924.

STAPLEDON, R. G. *The Land, Now and To-morrow*, London, 1935.

CHAPTER VIII

DEVELOPMENTS OF THE INDORE PROCESS

THE UTILIZATION OF TOWN WASTES

THE human population, for the most part concentrated in towns and villages, is maintained almost exclusively by the land. Apart from the harvest of the sea, agriculture provides the food of the people and the requirements of vegetable and animal origin needed by the factories of the urban areas. It follows that a large portion of the waste products of farming must be found in the towns and away from the fields which produced them. One of the consequences, therefore, of the concentration of the human population in small areas has been to separate, often by considerable distances, an important portion of the wastes of agriculture from the land. These wastes fall into two distinct groups:

(a) Town wastes consisting mainly of the contents of the dustbins, market, street, and trade wastes with a small amount of animal manure.

(b) The urine and faeces of the population.

In practically all cases in this country both groups of waste materials are treated as something to be got rid of as quickly, as unostentatiously, and as cheaply as possible. In Great Britain most town wastes are either buried in a controlled tip or burnt in an incinerator. Practically none of our urban waste finds its way back to the land. The wastes of the population, in most Western countries, are first diluted with large volumes of water and then after varying amounts of purification, are discharged either into rivers or into the sea. Beyond a little of the resulting sewage sludge the residues of the population are entirely lost to agriculture.

From the point of view of farming the towns have become parasites. They will last under the present system only as long as the earth's fertility lasts. Then the whole fabric of our civilization must collapse.

In considering how this unsatisfactory state of affairs can be remedied and how the wastes of urban areas can be restored to the soil, the magnitude of the problem and the difficulties which have to be overcome must be realized from the outset. These difficulties are of two kinds: those which belong to the subject proper, and those inherent in ourselves. The present system of sewage disposal has been the growth of a hundred years; problem after problem has had to be solved as it arose from the sole point of view of what seemed best for the town at the moment; mother earth has had few or no representatives on municipal councils to plead her cause; the disposal of waste has always been looked upon as the sole business of the town rather than something which concerns the well-being of the nation as a whole. The fragmentation of the subject into its urban components—medical, engineering, administrative, and financial—has followed; direction has been lost. The piecemeal consideration of such a matter could only lead to failure.

Can anything be done at this late hour by way of reform? Can mother earth secure even a partial restitution of her manurial rights? If the easiest road is first taken a great deal can be accomplished in a few years. The problem of getting the town wastes back into the land is not difficult. The task of demonstrating a working alternative to water-borne sewage and getting it adopted in practice is, however, stupendous. At the moment it is altogether outside the bounds of practical politics. Some catastrophe, such as a universal shortage of food followed by famine, or the necessity of spreading the urban population about the country-side to safeguard it from direct and indirect damage by hostile aircraft, will have to be upon us before such a question can even be considered.

The effective disposal of town wastes is, however, far less difficult, as will be seen by what has already been accomplished in this country. Passing over the earlier experiments with town wastes, summed up in a recent publication of the Ministry of Agriculture,[1] in which the dustbin refuse was used without modification, the recent results obtained with pulverized wastes, prepared by passing the sorted material (to remove tin cans, bottles, and other refractory objects) through a hammer mill, point clearly to the true role of this material in agriculture. Its value lies, not in its chemical composition, which is almost negligible, but in the fact that it is

---

[1] *Manures and Manuring*, Bulletin 36, Ministry of Agriculture and Fisheries, H.M. Stationery Office, 1937.

a perfect diluent for the manure heap, the weakest link in agriculture in many countries. The ordinary manure heap on a farm is biologically unbalanced and chemically unstable. It is unbalanced because the micro-organisms which are trying to synthesize humus have far too much urine and dung and far too little cellulose and lignin and insufficient air to begin with. It is unstable because it cannot hold itself together; the valuable nitrogen is lost either as ammonia or as free nitrogen; the micro-organisms cannot use up the urine fast enough before it runs to waste; the proteins are used as a source of oxygen with the liberation of free nitrogen. The fungi and bacteria of the manure heap are working under impossible conditions. They live a life of constant frustration which can only be avoided by giving them a balanced ration. This can be achieved by diluting the existing manure heaps with three volumes of pulverized town wastes. The micro-organisms are then provided with all the cellulose and lignin they need. The dilution of the manure heap automatically improves the aeration. The volume of the resulting manure is multiplied by at least three; its efficiency is also increased.

Such a reform of the manure heap is practicable. Two examples may be quoted. At the large hop garden at Bodiam in Sussex, the property of Messrs. Arthur Guinness, Son & Co., Ltd., over 30 tons of pulverized town wastes from Southwark are used daily throughout the year for humus manufacture. This material is railed in 6-ton truck-loads to Bodiam, transferred to the hop gardens by lorry and then composted with all the wastes of the garden—hop bine, hop string, hedge and roadside trimmings, old straw, all the farm-yard manure which is available—and every other vegetable and animal waste that can be collected locally. The annual output of finished humus is over 10,000 tons, which is prepared at an all-in cost of 10s. a ton, including spreading on the land. The Manager of this garden, Mr. L. P. Haynes, has worked out comparative figures of cost between nitrogen, phosphorus, and potash applied in the form of humus or artificials. The cost of town wastes for Bodiam is 4s. 6d. a ton; lorry transport from rail to garden 3s. a ton; assembling and turning the compost heaps and spreading on the land 2s. 6d. a ton. The analysis of this humus was: 0·96 per cent. nitrogen, 2·45 per cent. phosphate, and 0·62 per cent. potash. Sixteen tons of humus therefore contain 344 lb. of nitrogen, 769 lb. of P_2O_5, and 222 lb. of K_2O. The cost of this at 10s. a ton including spreading comes to £8 an acre. The purchase, haulage, and sowing of these amounts of NPK in the form of sulphate of ammonia, basic slag, and muriate of potash comes to £9. 12s. 7½d. There is therefore a distinct saving when humus is used. This, however, is only a minor item on the credit side. The texture of the soil is rapidly improving, soil fertility is being built up, the need for chemical manures and poison sprays to control pests is becoming less.

The manurial policy adopted on this hop garden has been confirmed in rather an interesting fashion. Before a serious attempt was made to prepare humus on the present scale, a small amount of pulverized Southwark refuse had been in use. The bulk of the manure used, however, was artificials supplemented by the various organic manures and fertilizers on the market. The labourers employed at Bodiam were therefore conversant with practically every type of inorganic and organic manure. One of their privileges is a supply of manure for their gardens. They have always selected pulverized town wastes because they consider this grows the best vegetables.

A second large-scale demonstration of the benefits which follow the reform of the manure heap has been carried out at Marden Park in Surrey. Many thousands of tons of humus have been made by composting pulverized town wastes with ordinary dung. In a paper read to the Farmers' Club on January 30th, 1939, Sir Bernard Greenwell refers to these results as follows: 'I have only two years' experience of this myself, but from the results I have seen we can multiply our dung by four and get crops as good as if the land had been manured with pure dung.' In 1938 I saw some of this work. Many of the fields on the estate had been divided into half, one portion being manured with humus and the other with an equal number of cartloads of dung. I inspected a number of these fields just as the corn was coming into ear. In every case the crops grown with humus—wheat, beans, oats, clover, and so forth—were definitely better than those raised with farm-yard manure. These results showed that this land wants freshly prepared humus, not so many lb. to the acre of this and that. In manuring we are nourishing a complex biological system not ministering to the needs of a conveyor belt in a factory.

Once the correct use of Southwark wastes was demonstrated a demand for this material arose. The sales increased; the demand now exceeds the supply. The details are given in Table 7.

TABLE 7

Sales of crushed wastes at Southwark

Year	Tons crushed*		Tons sold		Income from sales		
	T.	C.	T.	C.	£	s.	d.
	tons.	cwt.	tons.	cwt.	£	s.	d.
1933–4 . . .	18,643	12	7,971	9	653	9	9
1934–5 . . .	18,620	1	6,341	9	482	2	7
1935–6 . . .	19,153	14	9,878	5	1,001	11	1
1936–7 . . .	18,356	13	12,760	15	1,845	6	8
1937–8 . . .	18,545	15	15,391	8	2,306	13	7
1938–9 . . .	17,966	3	17,052	1	2,715	14	8

* A certain amount of these wastes is required by the Depot itself for sealing one of its own tips; so it is not possible to sell all the waste crushed to farmers.

When it is remembered that the annual dustbin refuse in Great Britain is in the neighbourhood of 13,000,000 tons and that about half of this material can be used for making the most of the urine and dung of our live stock, it will be evident what enormous possibilities exist for raising the fertility of the zones of land within, say, fifty miles of the large cities and towns. A perusal of the Public Cleansing Return for the year ending March 31st, 1938, published by the Ministry of Health, shows that a certain proportion of this dustbin refuse is still burnt in incinerators. Once, however, the agricultural value of this material is realized by farmers and market gardeners it will not be long before incineration is given up and the whole of the organic matter in our town wastes finds its way into the manure heap. When this time comes the utilization of the enormous dumps of similar wastes, which accumulated before controlled tipping was adopted, can be taken in hand. These contain many more millions of tons of material which can be dealt with on Southwark lines. In this way the manure heaps of a very large portion of rural England can be reformed and the fertility of a considerable area restored. A good beginning will then have been made in the restitution of the manurial rights owing to the country-side. The towns will have begun to repay their debt to the soil.

Besides the wastes of the dustbins and the dumps there is another and even more important source of unused humus in the neighbourhood of our cities and towns. This occurs in the controlled tips in which most of the dustbin refuse is now buried. In controlled tipping the town wastes are deposited in suitable areas near cities and sealed with a layer of clay, soil, or ashes so as to prevent nuisance generally and also the breeding of flies. The seal, however, permits sufficient aeration for the first stage in the conversion of most of the organic matter into humus. The result is that in a year or two the tip becomes a humus mine. The crude organic matter in these wastes is slowly transformed by means of fungi and bacteria into humus. All that is needed is to separate the finely divided humus from the refractory material and to apply it to the land.

A very valuable piece of research work on this matter has recently been undertaken at Manchester. The results are described by Messrs. Jones and Owen in *Some Notes on the Scientific Aspects of Controlled Tipping*, published by the City of Manchester. The main object of the work was to establish the facts underlying controlled tipping so that any discussion on the efficacy of this process, as compared with incineration, could be conducted on the basis of carefully ascertained knowledge. The investigation, however, is invaluable from the agricultural standpoint. The experiments were begun in August 1932 at Wythenshawe in a controlled tip on a piece of low-lying marshy ground subject to periodic flooding from the adjacent river Mersey. One of the subsidiary objects of the tipping was to reclaim the land for recreational or other uses in the future. Six experimental plots were selected for the tests, each approximately 16 feet by 12 feet. The material contained in the tip was ordinary dustbin refuse tipped to a depth of 6 feet. The first object was to ascertain the consequences of bacterial action on the organic matter in the interior of the tip, such as the generation of temperature, the biological as well as the chemical changes, and any alteration in the gaseous atmosphere in the interior of the mass. Having disposed of these preliminary matters, it was proposed to attack the main problem and to answer the question: Is controlled tipping safe?

Careful attention was first given to the seal. The surface of the plots was covered with a layer of fine dust and ashes, of a minimum thickness of 6 inches, obtained by passing household refuse over a ⅜ inch mesh. Such a seal, which contained about 2·5 per cent. of organic matter, proved to be a suitable mechanical covering and also prevented the breeding of flies. The sides and ends of the experimental plots were covered with clay well tamped down. The plots therefore behaved as if they were large flowerpots in direct contact with the moist earth below but separated from the outer atmosphere by a permeable seal of screened dust and fine ashes.

The unsorted household refuse under experiment represented an average sample and contained about 42 per cent. of organic matter, the remaining 58 per cent. being composed of inorganic materials. After tipping and sealing, there was a rapid rise of temperature, irrespective of the season, to a maximum of 160° F. towards the end of the first week. This was caused by the activities of the thermogenic and thermophyllic members of the aerobic group of bacteria which break down cellulose, liberate heat, and produce large volumes of carbon dioxide. At the same time these organisms rapidly multiply and in so doing synthesize large amounts of protein from the mixed wastes. This on the death of the organisms forms a valuable constituent of the humus left when the bacterial activities die down after about fifteen weeks, as is indicated by the return of the temperature of the tip to normal. The controlled tip therefore behaves very much like an Indore compost heap.

As would be expected from the heterogeneous nature and uneven distribution of the contents of the tip, considerable variations were shown in the maximum temperatures attained. During the period of fermentation the bacterial flora (at first aerobic) use and reduce the oxygen content of the tip, and so pave the way for the facultative anaerobic organisms which complete the conversion of the organic matter into humus.

A detailed examination of the gases produced in the tips showed that in addition to nitrogen, carbon dioxide, and oxygen, a considerable quantity of methane (16 per cent.) and smaller proportions of carbon monoxide (2·8 per cent.) and hydrogen (2·5 per cent.) occurred. Traces only of sulphuretted hydrogen were detected. The presence of carbon monoxide, methane, and hydrogen would naturally result from the anaerobic fermentation which establishes itself in the second stage of the production of humus after the free oxygen in the tip becomes exhausted. These gases are similar to those produced by the decay of organic matter in swamp rice cultivation in India, where the supply of oxygen is almost always in defect. The absence of anything beyond a trace of sulphuretted hydrogen is reassuring, as this proves that the intense reduction which precedes the formation of the salts of alkali soils does not occur in a controlled tip.

The manurial value of the humus in the tips was determined by analysis and valuation. The average content of nitrogen was 0·8 per cent., of phosphoric acid 0·5 per cent., of potash 0·3 per cent. The estimated value of the dry material per ton was 10s. This value, however, will have to be multiplied by a factor ranging from 2 to 2·5, because experience has shown that the market price of organic manures, based on supply and demand, is anything from two to two and a half times greater than that calculated from the chemical analysis. The unit system of valuation applies only to artificial manures like sulphate of ammonia made in factories; it does not hold in the case of natural manures like humus.

One of the last sections of the Report relates to the danger of infectious diseases as a possible consequence of controlled tipping. The authors conclude that 'danger arising from possible presence of pathogenic germs in a controlled tip may be dismissed as non-existent'.

One of the plots, No. 1, not only developed a high temperature but showed a much more gradual fall than the other plots. This was apparently due to the higher content of organic matter combined with better aeration. The results of this plot suggest that more and better humus might be obtained in a controlled tip if the object of tipping were, as it should be, to secure the largest amount of humus of the best possible quality. It would not be a difficult matter to increase the oxygen intake at the beginning by allowing more and more air to diffuse in from the atmosphere. This could perhaps be done most easily and cheaply by reducing the thickness of the seal by about a third. If the seal were reduced in this way, ample air would find its way into the fermenting mass in the early stages; the humus would be improved; the covering material saved could be used for a new seal. The controlled tip would then become a very efficient humus factory.

In countries where there is no system of water-borne sewage there has been no difficulty in converting the wastes of the population into humus. The first trials of the Indore Process for this purpose were completed in Central India in 1933 by Messrs. Jackson and Wad at three centres near Indore—the Indore Residency, Indore City, and the Malwa Bhil Corps. Their results were soon taken up by a number of the Central India and Rajputana States and by some of the municipalities in India. Subsequent developments of this work, including working drawings and figures of cost, were summed up in a paper read to the Health Congress of the Royal Sanitary Institute held at Portsmouth in 1938. This document has been reproduced as Appendix C (p. 235). A perusal of this statement shows that human wastes are an even better activator than animal residues. All that is necessary is to provide for abundant aeration in the early stages and to see that the night soil is spread in a thin film over the town wastes and that no pockets or definite layers are left. Both of these interfere with aeration, produce smell, and attract flies. Smell and flies are therefore a very useful means of control. If the work is properly done there is no smell, and flies are not attracted because the intense oxidation processes involved in the early stages of the synthesis of humus are set in motion. It is only when the air supply is cut off at this stage that putrefactive changes occur which produce nuisance and encourage flies.

Whether or not it will always be necessary to erect permanent installations for converting night soil and town refuse into humus, experience only can decide. In a number of cases it may be easier to do the composting daily in suitable pits or trenches on the lines described on p. 240. In this way the pits or trenches themselves become temporary composting chambers; no turning is required; the line of pits or trenches can soon be used for agricultural purposes—for growing all kinds of fodder, cereal, and vegetable crops. At the same time the land is left in a high state of fertility.

A number of medical officers all over the world are trying out the composting of night soil on the lines suggested. In a few years a great deal of experience will be available, on which the projects of the future can be based.

As far as countries like Great Britain are concerned, the only openings for the composting of night soil occur in the country-side and in the outer urban zones where the houses are provided with kitchen gardens. In such areas the vast quantities of humus in the controlled tips can be used in earth-closets and the mixed night soil and humus can be lightly buried in the gardens on the lines so successfully carried out by the late Dr. Poore and described in his *Rural Hygiene*, the second edition of which was published in 1894.

Since Dr. Poore's work appeared a new development in housing has taken place in the garden cities and in colonies like those started by the Land Settlement Association. Here, although there is ample land for converting every possible waste into humus, the water-borne method of sewage disposal and the dust-carts of the crowded town have been slavishly copied. In an interesting paper published in the *British Medical Journal* of February 9th, 1924, Dr. L. J. Picton, then Medical Officer of Health of the Winsford Urban District, Cheshire, pointed out how easy it would be to apply Dr. Poore's principles to a garden city.

'A plot of 4 acres should be taken on the outskirts of a town and twenty houses built upon it. Suppose the plot roughly square, and the road to skirt one corner of it. Then this corner alone will possess that valuable quality "frontage". Sacrifice this scrap of frontage by making a short gravelled drive through it, to end blindly in a "turn-round" in the middle of the plot. The houses should all face south—that is to say, all their living rooms should face south. They must therefore be oblong, with their long axes east and west (Fig. 1). The larder, the lobby, lavatory, staircase and landing will occupy the north side of each house. The earth closet is best detached but approached under cover—a cross-ventilated passage or short veranda, or, if upstairs, a covered bridge giving access to it. The houses should be set upon the plot in a diamond-shaped pattern, or in other words, a square with its corners to north, south, east and west. Thus one house will occupy the northernmost point of the plot, and from it, to the south-east and south-west, will run a row of some five or six houses a side, arranged in echelon. Just as platoons in echelon do not block each other's line of

I

fire so houses thus arranged will not block each other's sunlight. A
dozen more houses echeloned in a V with its apex to the south will
complete the diamond-shaped lay-out. The whole plot would be
treated as one garden, and one whole-time head gardener, with the

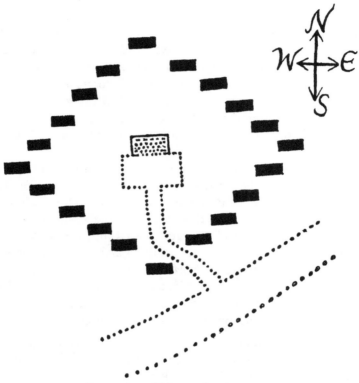

FIG. 1. A model layout for 20 cottages.

help he needed, would be responsible for its cultivation. The daily
removal of the closet earth and its use as manure—its immediate
committal to the surface soil and its light covering therewith—would
naturally be amongst his duties. A gardener using manure of great
value, not a scavenger removing refuse; a "garden rate" paid by each
householder, an investment productive of fresh vegetables to be had at
his door, and in one way or another repaying him his outlay, not to
speak of the amenity added to his surroundings, instead of a "sanitary
rate" paid to be rid of rubbish—such are the bases of this scheme.'

What is needed are a few working examples of such a housing
scheme and a published account of the results. These, if successful,

would at once influence all future building schemes in country districts and would point the way to a considerable reduction in rents and rates. The garden-city and water-borne sewage are a contradiction in terms. Water-borne sewage has developed because of overcrowding and the absence of cultivated land. Remove overcrowding and the case for this wasteful system disappears. In the garden city there is no need to get rid of wastes by the expensive methods of the town. The soil will do it far more efficiently and at far less cost. At the same time the fertility of the garden city areas will be raised and large crops of fresh vegetables and fruit—one of the factors underlying health—will be automatically provided.

Such a reform in housing schemes will not stop at the outer fringes of our towns and cities. It will be certain to spread to the villages and to the country-side, where a few examples of cottage gardens, rendered fertile by the wastes of the inhabitants, are still to be found here and there. More are needed. More will arise the moment it is realized that the proper utilization of the wastes of the population depends on composting processes and the correct use of humus. All the trouble, all the expense, and all the difficulties in dealing with human wastes arise from following the wrong principle—water—and setting in motion a vast train of putrefactive processes. The principle that must be followed is abundant aeration at the beginning: the conversion of wastes into humus by the processes Nature employs in every wood and every forest.

BIBLIOGRAPHY

GREENWELL, SIR BERNARD. 'Soil Fertility: the Farm's Capital', *Journal of the Farmers' Club*, 1939, p. 1.

HOWARD, SIR ALBERT. 'Preservation of Domestic Wastes for Use on the Land', *Journal of the Institution of Sanitary Engineers*, xliii, 1939, p. 173.

—— 'Experiments with Pulverized Refuse as a Humus-Forming Agent', *Journal of the Institute of Public Cleansing*, xxix, 1939, p. 504.

JONES, B. B., and OWEN, F. *Some Notes on the Scientific Aspects of Controlled Tipping*, City of Manchester, 1934.

PICTON, L. J. 'The Economic Disposal of Excreta: Garden Sanitation', *British Medical Journal*, February 9th, 1924.

POORE, G. V. *Essays on Rural Hygiene*, London, 1894.

Public Cleansing Costing Returns for the year ended March 31st, 1938, H. M. Stationery Office, 1939.

PART III

HEALTH, INDISPOSITION, AND DISEASE IN AGRICULTURE

CHAPTER IX

SOIL AERATION

THE transformation of soil fertility into a crop is only possible by means of oxidation processes. The various soil organisms— bacteria and fungi in particular—as well as the active roots need a constant supply of oxygen. As soon as this was recognized, aeration became an important factor in the study of the soil. In this matter, however, practice has long preceded theory: many devices such as sub-soil drainage, sub-soiling, as well as mixed cropping— all of which assist the ventilation of the soil—have been in use for a long time.

The full significance of soil aeration in agriculture has only been recognized by investigators during the last quarter of a century. The reason is interesting. Till recent years most of the agricultural experiment stations were situated in humid regions where the rainfall is well distributed. Rain is a saturated solution of oxygen and is very effective in supplying this gas to the soil whenever percolation is possible. Hence in such regions crops are not likely to suffer from poor aeration to anything like the same extent as those grown in the arid regions of North-West India where the soils are silt-like and most of the moisture has to be supplied by irrigation water low in dissolved oxygen. Such soils lose their porosity with the greatest ease when flooded; the minute particles run together and form an impermeable surface crust. Only when the humus content is kept high can adequate permeability be maintained. Long before the advent of the modern canal, the cultivators of India had acted on this principle. The organic matter content of the areas commanded by wells has always been maintained at a high level. Irrigation engineers and Agricultural Departments have been slow to utilize this experience. Canal water has been provided, but no steps have been taken simultaneously to increase the humus content of the soil.

PLATE III

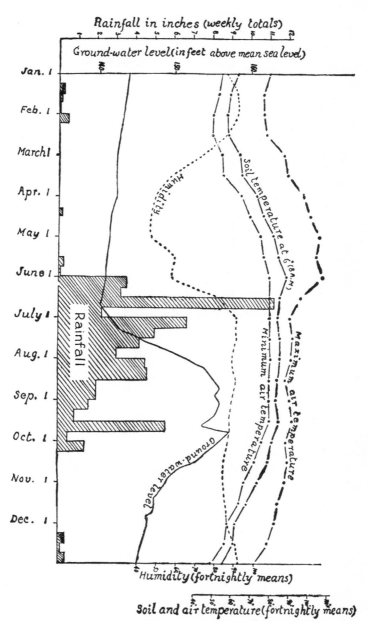

RAINFALL, TEMPERATURE, HUMIDITY, AND DRAINAGE,
PUSA 1922

It follows from the constant demands of the soil for fresh air that any agency which interferes, even partially or temporarily, with aeration must be of supreme importance in agriculture. A number of factors occur which bring about every gradation between a restricted oxygen supply and complete asphyxiation. The former result in infertility, the latter in the death of the soil.

How does the plant respond to soil conditions in which oxygen becomes the limiting factor? Generally speaking there is an immediate reaction on the part of the root system. This is well seen in forest trees and in the undergrowth met with in woodlands. The roots adjust themselves to the new conditions; the trees establish themselves and at the same time improve the aeration and also add to the fertility of the soil; incidentally all other competitors are vanquished. Soil aeration cannot therefore be studied as if it were an isolated factor in soil science. It must be considered along with (1) the responses of the root system to deficient air, (2) the relation between root activity and soil conditions throughout the year, and (3) the competition between the roots of various species. In this way the full significance of this factor in agriculture and in the maintenance of soil fertility becomes apparent. This is the theme of the present chapter. An attempt will be made to explain soil aeration as it affects the plant in relation to the environment and to show how the plant itself can be used as a research agent.

THE SOIL AERATION FACTOR IN RELATION TO GRASS AND TREES

Between the years 1914 and 1924 the factors involved in the competition between grass and trees were investigated by me at Pusa. Three main problems were kept in view, namely, (1) why grass can be so injurious to fruit trees, (2) the nature of the weapons by which forest trees vanquish grass, and (3) the reaction of the root system of trees to the aeration of the soil. An account of this study was published in the *Proceedings of the Royal Society of London* in 1925 (B, vol. xcvii, pp. 284–321). As the results support the view that in the investigation of the soil aeration factor the plant can always make an important contribution, a summary of the main results and a number of the original illustrations have been included in this chapter.

The climatic factors at Pusa are summed up in Plate III. It will be seen that after the break of the south-west monsoon in June, the humidity rises followed by a steady upward movement in the ground water-level till October when it falls again. In 1922 the total rise of the sub-soil water-level was 16·5 feet, a factor which is bound to interfere with the oxygen supply, as the soil air which is rich in carbon dioxide is slowly forced into the atmosphere by the ascending water-table.

The soil is a highly calcareous silt-like loam containing about 75 per cent. of fine sand and about 2 per cent. of clay. About 98 per cent. will pass through a sieve of 80 meshes to the linear inch. There is no line of demarcation between soil and sub-soil: the sub-soil resembles the soil and consists of alternating layers of loam, clay, and fine sand down to the sub-soil water, which normally occurs about 20 feet from the surface. The percentage of calcium carbonate is often over 30, while the available phosphate is in the neighbourhood of 0·001 per cent. In spite of this low content of phosphate, the tract in which Pusa is situated is highly fertile, maintaining a population of over 1,200 to the square mile and exporting large quantities of seeds, tobacco, cattle, and surplus labour without the aid of any phosphatic manures. The facts relating to agricultural production in this tract flatly contradict one of the theories of agricultural science, namely, the need for phosphatic fertilizers in areas where soil analysis shows a marked deficiency in this element. Two other factors, however, limit crop production—shortage of humus and loss of permeability during the late rains due to a colloidal condition of the soil; the pore spaces near the surface become water-logged; percolation stops and the soil is almost asphyxiated, a condition which is first indicated by the behaviour of the root system and then by restricted growth.

For the investigation of the soil aeration factor in relation to grass and trees at Pusa, eight species of fruit trees—three deciduous and five evergreen—were planted out in three acres of uniform land, each species being raised from a single parent. The plan (Fig. 2) gives further details and makes the arrangement clear. Two years after planting, when the trees were fully established and remarkably even, a strip including nine trees of each of the eight rows was laid down to grass. The two end plots, which were clean cultivated, served as controls. When the grass was well established and its injurious effect on the young trees was clearly marked, the three southern trees of the grass plot were provided with aeration trenches, 18 inches wide and 24 inches deep filled with broken bricks, these trenches being made midway between the lines of trees. To ascertain the effect of grass on established

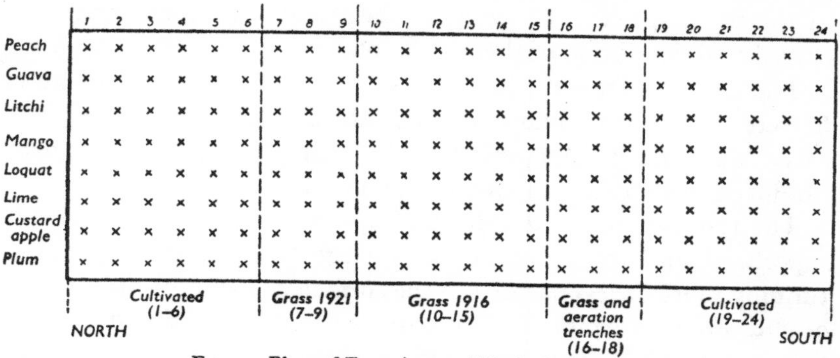

	1	2	3	4	5	6	7	8	9	10	11	12	13	14	15	16	17	18	19	20	21	22	23	24
Peach	×	×	×	×	×	×	×	×	×	×	×	×	×	×	×	×	×	×	×	×	×	×	×	×
Guava	×	×	×	×	×	×	×	×	×	×	×	×	×	×	×	×	×	×	×	×	×	×	×	×
Litchi	×	×	×	×	×	×	×	×	×	×	×	×	×	×	×	×	×	×	×	×	×	×	×	×
Mango	×	×	×	×	×	×	×	×	×	×	×	×	×	×	×	×	×	×	×	×	×	×	×	×
Loquat	×	×	×	×	×	×	×	×	×	×	×	×	×	×	×	×	×	×	×	×	×	×	×	×
Lime	×	×	×	×	×	×	×	×	×	×	×	×	×	×	×	×	×	×	×	×	×	×	×	×
Custard apple	×	×	×	×	×	×	×	×	×	×	×	×	×	×	×	×	×	×	×	×	×	×	×	×
Plum	×	×	×	×	×	×	×	×	×	×	×	×	×	×	×	×	×	×	×	×	×	×	×	×

Cultivated (1–6) — NORTH | Grass 1921 (7–9) | Grass 1916 (10–15) | Grass and aeration trenches (16–18) | Cultivated (19–24) — SOUTH

FIG. 2. Plan of Experimental Fruit Area, Pusa.

trees in full bearing, the southern strip of the northern control plot was grassed over in 1921. The general results of the experiment, as seen in 1923, are shown in Plate IV. The harmful effect of grass on fruit trees at Pusa is even more intense than on clay soils like those of Woburn in Great Britain. Several species were destroyed altogether within a few years.

As great differences in root development were observed between the trees under grass, under grass with aeration trenches, and under clean cultivation, the first step in investigating the cause of the harmful effect of grass appeared to be a systematic exploration of the root system under clean cultivation so as to establish the general facts of distribution, to ascertain the regions of root activity during the year and to correlate this information with the growth of the above ground portion of the trees. This was carried out in 1921 and the work was repeated in 1922 and again in 1923. The method adopted was direct: to expose the root system quickly and to use a fine water-jet for freeing the active roots from the soil particles. By using a fresh tree for each examination and by employing relays of labourers, it was possible to expose any desired portion of the root system down to 20 feet in a few hours and to make the observations before the roots could react to the new conditions.

THE ROOT SYSTEM OF DECIDUOUS TREES

The root systems of three deciduous trees—the plum, the peach, and the custard apple—were first studied. The results obtained in the three species were very similar, so it is only necessary to describe in detail one of them—the plum.

The local variety of plum sheds its leaves in November and flowers profusely in February and March. The fruit ripens in early May, the hottest period of the year. The new shoots are produced during the hot weather and early rains.

The root system is extensive and appears at first to be entirely superficial and to consist of many large freely branching roots running more or less parallel to the surface in the upper 18 inches of soil. Further exploration disclosed a second root system. From the under side of the large surface roots, smaller members are given off which grow vertically downwards to about 16 feet from the surface. These break up into many branches in the deep layers of moist fine sand, just above the water-table. The Indian variety of plum therefore has two root systems (Plate V, Fig. 1). The deep root system begins to develop soon after the young trees are planted out.[1]

During the resting period (December to January) occasional absorbing roots are formed in the superficial system. When flowering begins, the formation of new rootlets spreads from the surface to the deep soil layers. As the surface soil dries in March, the active roots on the superficial system turn brown and die and this portion passes into a dormant condition. From the middle of March to the break of the rains in June, root absorption is confined entirely to the deeper layers of soil. Thus on April 14th, 1921, when the trees were ripening their fruit and making new growth during a period of intense heat and dryness, most of the

---

[1] In August 1923 the root systems of young custard apples, mangoes, guavas, limes, and loquats, planted in March 1922 were examined. The young vertical roots varied in length from 10 inches in the custard apple and lime to 1 foot in the mango, 1 foot 2·5 inches in the guava and 1 foot 8 inches in the loquat. Newly planted trees form the superficial system first of all, followed rapidly by the deep system.

PLATE IV

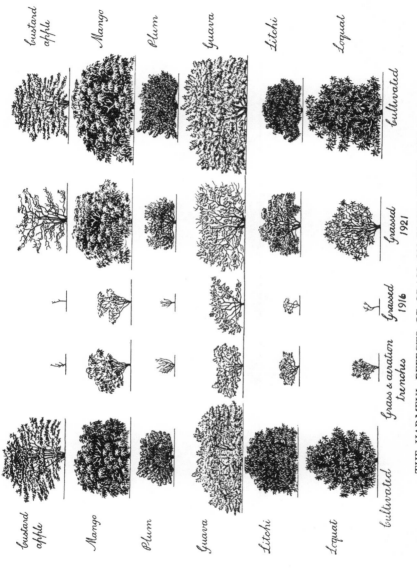

THE HARMFUL EFFECTS OF GRASS ON FRUIT TREES, PUSA, 1923

water, nitrogen, and minerals necessary for growth were absorbed from a layer of moist fine sand between 10 feet 6 inches and 15 feet below the surface. This state of affairs continues till the break of the rains in June when a sudden change takes place. The moistening of the surface soil rapidly brings the superficial root system into intense activity. These hitherto dormant roots literally break into new active rootlets in all directions, the process beginning about thirty hours after the first fall of rain. In the early monsoon therefore the trees use the whole of the root system, both superficial and deep. A change takes place during late July as the level of the ground water rises. In early August active roots are practically confined to the upper 2 feet of soil. Absorption is now restricted to the surface system. At this period the active roots react to the poor soil aeration due to the rise in the ground water-level by growing towards the atmosphere and even out of the soil into the air, particularly under the shade of the trees and where the soil is covered by a layer of dead leaves (Plate V, Fig. 3). This aerotropism continues till early October, when the growth above ground stops and the trees ripen their wood preparatory to leaf fall and the cold weather rest. During October, as the level of the ground water falls and air is drawn into the soil, there is some renewal of root activity near the surface and down to 3 feet.

One interesting exception to this periodicity in the root activity of the plum occurs. Falls of rain, nearly an inch in amount, sometimes occur during the hot season. The effect on the superficial root system of the plum of three of these storms was investigated. When the rainfall was 0·75 of an inch or more, the surface roots at once responded and produced a multitude of new absorbing roots. As the soil dried these ceased to function and died. In one case, where the rainfall was only 0·23 inches, no effect was produced. Irrigation during the hot weather acts in a similar manner to these sudden falls of rain. It maintains the surface root system in action during this period and explains why irrigation during the hot months is necessary on the alluvium if really good quality fruit is to be obtained. It is true that without artificial watering the trees ripen a crop at Pusa, but in size and quality the crop is greatly inferior to that obtained with the help of irrigation. Either root system will produce a plum. High quality is obtained only when the surface system functions; poor quality always results when the deep system only is in action.

In the detailed examination of the active surface roots of the plum and of the seven other species in this experiment, fresh fungous mycelium was often observed running from the soil towards the growing roots. In the deeper soil layers this was never observed. In all probability this mycelium is connected with the mycorrhizal association so common in fruit trees. This matter was not carried further at the time. It is, however, more than probable that all the eight species of fruit trees in the Pusa Experiment are mycorrhiza-formers and that the fungus observed round the active roots was concerned with this association. The mycorrhizal relationship in the surface roots is probably involved in the production of high quality fruit. Plants with two root systems such as these are therefore admirably adapted for the future study of the relation between humus in the soil, the mycorrhizal association, and the development of quality. It would not be difficult to compare plants grown side by side on the sub-soil (to remove the humus occurring in the surface soil), the one manured with complete artificials, the other with freshly prepared humus. In the former there would be little or no mycorrhizal invasion; in the latter it would probably be considerable. If, as is most likely, the mycorrhizal association enables the tree to absorb nutrients in the organic form by the digestion of fungous mycelium, this would explain why quality only results when the surface roots are in action.

Support for the view of plant nutrition suggested in the preceding paragraph was supplied by the custard apple, the root development of which is similar to that of the plum and peach. In the custard apple new shoots are formed in the hot weather when the water, nitrogen, and nutrients are obtained from the deep soil layers only. After the break in the rains and the resumption of root activity on the surface, the leaves increase in size (from 5.8×2.6 cm. to 10.5×4.5 cm.), develop a deeper and healthier green, while the internodes lengthen (Fig. 3). The custard apple records the results of these various factors in the size and colour of its leaves and in this way acts as its own soil analyst.

While this book was being printed specimens of the young active roots of the custard apple, mango, and lime were collected in Mr. Hiralal's orchard, Tukoganj, Indore, Central India, on November 11th, 1939, by Mr. Y. D. Wad. They were examined by Dr. Ida Levisohn on December 19th, 1939, who reported that all

FIG. 3. Hot weather (below *aa*) and monsoon foliage (above *aa*) of the custard apple.

three species showed typical endotrophic mycorrhizal infection indicated macroscopically by the absence of root hairs, or great reduction in their number, and, in the mango particularly, by beading. The active hyphae in all three cases were of large diameter, with thin walls and granular contents, the digestion stages occurring in the inner cortex with clumping of mycelium, remains of hyphae and homogeneous granular masses. Absorption of the fungus appeared to be taking place with great rapidity. In the custard apple the same kind of mycelium was found outside the roots and connected with them.

THE ROOT SYSTEM OF EVERGREENS

The most interesting root system of the five evergreens studied—mango, guava, litchi, sour lime, and loquat—was the guava.

The guava drops its foliage in early March, simultaneously producing new leaves. It proved an excellent plant for the study of the root system, as the reddish roots are strongly developed and easy to follow in a grey alluvial soil like that of Pusa. There is an abundant superficial system giving off numerous branches which grow downwards to the level of permanent water (Plate VI, Fig. 1). The whole of the root system, superficial and deep, was found to be active at the beginning of the hot weather (March 21st, 1921), the chief zone of activity occurring in a moist layer of fine sand 10 feet 4 inches to 14 feet 7 inches from the surface. As the hot weather became established, the absorbing roots of the guava near the surface dried up and root activity was confined to the deeper layers of soil. In 1922 the monsoon started on June 3rd. An exposure of the surface roots was made on June 5th, forty-eight hours after the rains started. From 1 foot 5 inches to 12 feet new roots were found in large numbers, the longest measuring 1 cm. As the soil became moistened by the early rains, the dormant zone produced new roots from above downwards till the whole root system became active. After July a change takes place as the ground water rises, the deep roots becoming dormant as immersion proceeds. On August 25th, 1922, root activity was mainly confined to the surface system in the upper 29 inches of soil, the last active root occurring at 40 inches. In the late rains the active roots escape asphyxiation by becoming strongly aerotropic (Plate VI, Fig. 4). An interesting change takes place after the level of the sub-soil water falls in October and the aeration of the lower soil layers is renewed. The deep root system again becomes active in November, the degree of activity depending on the monsoon rainfall (Plate VI, Fig. 5). In 1921, a year of short rainfall when the rise of the ground water was very small, the deep roots came into activity in November down to 15 feet 3 inches. The next year—November 1922—when the monsoon and the rise of the ground water were both normal, root activity did not extend below 5 feet 7 inches.

Although the guava is able to make new growth during the hot season by means of its deep root system it is a decided advantage if the surface roots are maintained in action by means of irrigation. Surface watering in the hot weather of 1921 increased the size of the leaves from 9·1 × 4·0 cm. to 11·6 × 5·0 cm. and greatly improved their colour.

The root system and the development of active roots in the mango, litchi, lime, and loquat follow generally what has been described in the guava. All these species give off vertical roots from the surface system, but in the case of the litchi and the lime these did not penetrate to the deeper layers. The roots of all four species exhibit marked aerotropism in the late rains. The vertical roots of the lime were always unable to penetrate the deeper layers of clay.

THE HARMFUL EFFECT OF GRASS

The harmful effect of grass on fruit trees varies with the species and with the period in the life of the tree when the grass is planted. Young trees are more adversely affected than fully developed individuals, which contain large quantities of reserves in the wood. Deciduous species suffer more than evergreens. These facts suggest that the harmful effect of grass is a consequence of starvation.

The effect of grass on young trees was first studied. The custard apple was the most sensitive. The trees were killed in 1916 within the first two years after the grass was planted. Next in order of susceptibility were the loquat (all died before the end of 1919), the plum, the lime, and the peach. The litchi and the mango just managed to maintain themselves. The guava was by far the least affected, the trees under grass being almost half the height of those under clean cultivation.

Grass not only reduces the amount of new growth but affects the leaves, branches, old wood, and fruit as well as the root system. The results relating to the above ground portion of the trees closely follow those described by the Woburn investigators. Compared with the foliage produced under clean cultivation, the leaves from the trees under grass appear later, are smaller and yellower and fall prematurely. The internodes are short. The bark of the twigs is light coloured, dull, and unhealthy and quite different from that of healthy trees. The bark of the old wood has a similar appearance and attracts lichens and algae to a much greater extent than that of the cultivated trees. The trees under grass flower late and sparingly. The fruit is small, tough, very highly coloured, and ripens earlier than the normal.

The effect of grass on the root system is equally striking. Except in the guava, the effect of grass on the superficial system is to restrict the amount of root development, to force the roots below the grass, and to reduce the number of active roots during the monsoon. The guava is an exception. The surface system is well developed, the roots are not driven downwards by the grass while active rootlets are readily formed in the upper 4 inches of soil soon after the rains begin, very much as in the cultivated trees. In August 1922, when the ground water had risen to its highest point, the absorbing roots of the guava were found in the surface film of soil, and also above the surface among the stems of the grass. The grass carpet therefore acts as an asphyxiating agency in all these species, the guava excepted.

The grass covering has no appreciable effect either on the development or on the activity of the deep roots. This portion of the root system was explored during the hot weather of 1921 in the case of the guava (Plate VI, Fig. 3), mango, and litchi and results were obtained very similar to those in the corresponding cultivated trees.

Grass not only affects the roots underneath but also the development of those of the neighbouring trees under cultivation. Such roots either turn away from the grass, as in the custard apple, or else turn sharply downwards before they reach it.

A number of conclusions can be drawn from these root exposures. The custard apple, loquat, peach, and lime are unable to maintain their surface root systems under grass, but behave normally as regards the deep root system. Only the guava is able to get its roots above those of the grass during the rains.

The study of the harmful effect of grass on established trees also yielded interesting results. In this case the trees carried ample reserves in the wood and, as might be expected, the damage was less spectacular than in the case of young trees with little or no reserves. The order of susceptibility to grass, however, was very much the same in the two cases. When the fully-grown trees were first put under grass in August 1921, the grass at first grew poorly in tufts with bare ground between. Even this imperfect covering soon affected the custard apples, loquats, peaches, and litchis. By the rains of 1922 the grass became continuous; the effect on the trees was then much more marked.

In the plum interesting changes occurred. In July 1922, less than a year after planting the grass, the new shoots showed arrested growth and the foliage was attacked by leaf-destroying insects, which, however, ignored the leaves of the neighbouring cultivated plot. If the insects were the real cause of the trouble, it is difficult to see why the infection did not spread beyond the trees under grass. In January 1923 the average length of the new wood in these trees was 1 foot 5 inches compared with 3 feet 7 inches in the controls. The twigs were dull and purplish, the internodes were short (Plate V, Fig. 5). In February 1923 flowering was restricted and in April only tufts of leaves were formed at the ends of the branches instead of new shoots (Plate V, Fig. 8). Early in 1924, when I left Pusa and had to discontinue the work, a great deal of die-back was taking place.

Very similar results were obtained in all the species except the mango, which resisted grass better than any of the others. No definite effect was observed in this species till June 1923, when the foliage became distinctly lighter than that of the cultivated trees. The general results brought about by grass in all these cases suggested that the trees were slowly dying from starvation.

A year after the grass was planted and the grass effect was becoming marked, the root system of these established trees was examined. In August 1922 the plums, peaches, custard apples, mangoes, litchis, and loquats under grass were found to have produced very few active rootlets in the upper foot of soil compared with the controls. In the case of the custard apples and the loquats, which suffered most from grass, there was a marked tendency for the new roots to grow downwards and away from the grass. No differences were observed in the dormancy or activity of the deep root system as compared with the controls. The deep roots behaved exactly like those under clean cultivation.

During these examinations two instances of the striking effect of increased aeration on root development were observed. In July 1923 burrowing rats took up their quarters under one of the limes and one of the loquats, in each case on the southern side. Shortly afterwards the leaves just above the rat holes became very much darker in colour than the rest. Examination of the soil

FIG. 4. The effect of burrowing rats on the growth of the plum under grass (June 21st, 1923).

immediately round the burrows showed a copious development of new active rootlets, far greater even than in the surface soil of the cultivated plot. The extra aeration had a wonderfully stimulating effect on the development of active roots, even under grass. The appearance of the leaves suggested an application of nitrogenous manure. Similar observations were made in the case of the plum (Fig. 4). Here the burrows caused a dying tree to produce new growth.

THE EFFECT OF AERATION TRENCHES ON YOUNG TREES UNDER GRASS

The effect of aeration trenches in modifying the influence of grass suggests that one of the factors at work is soil asphyxiation. In the case of the custard apple and the lime the aeration trenches had no effect; all the trees died. The death of the plums was delayed by the aeration trenches. The loquats, litchis, and man-

PLATE V

PLUM (*Prunus communis*, Huds.).

FIG. 1. Superficial and deep roots (April 25, 1921).
FIG. 2. The repair of the deep root-system (August 6, 1923).
FIG. 3. Superficial rootlets growing towards the surface (August 12, 1922).
FIGS. 4 and 5. New wood under cultivation and grass (January 25, 1923).
FIGS. 6 and 7. New shoots and leaves under clean cultivation (April 5, 1923).
FIGS. 8 and 9. The corresponding growth under grass (April 5, 1923).

goes benefited considerably. In the guavas the trees provided with aeration trenches were indistinguishable from those under grass. The general results are shown in Table 8, in which the measurements of a hundred fully-developed leaves, made in March 1921, are recorded.

At the end of 1920 the roots were exposed to a depth of 2 feet in order to ascertain the effect of the extra aeration on the development of the superficial system. The results were interesting. In all cases the superficial roots were much larger and better developed

TABLE 8

The reduction in leaf size under grass

	Grass	Grass with aeration trenches	Cultivated
	cm.	cm.	cm.
Plum	$3 \cdot 2 \times 1 \cdot 1$	$4 \cdot 6 \times 1 \cdot 7$	$7 \cdot 1 \times 2 \cdot 9$
Peach	$7 \cdot 1 \times 1 \cdot 8$	$8 \cdot 4 \times 2 \cdot 3$	$11 \cdot 4 \times 3 \cdot 1$
Guava	$8 \cdot 1 \times 3 \cdot 2$	$10 \cdot 6 \times 4 \cdot 4$	$11 \cdot 3 \times 4 \cdot 4$
Mango	$11 \cdot 2 \times 2 \cdot 9$	$13 \cdot 7 \times 3 \cdot 8$	$20 \cdot 9 \times 5 \cdot 5$
Litchi	$8 \cdot 9 \times 2 \cdot 4$	$11 \cdot 5 \times 3 \cdot 4$	$12 \cdot 2 \times 3 \cdot 5$
Lime	$3 \cdot 8 \times 1 \cdot 6$	$5 \cdot 2 \times 2 \cdot 1$	$6 \cdot 4 \times 3 \cdot 4$
Loquat	Trees dead	$16 \cdot 4 \times 4 \cdot 6$	$22 \cdot 1 \times 5 \cdot 9$

than those under grass, except in the guava where no differences in size could be detected. The roots were attracted by the trenches, often branching considerably in the soil at the side of the trenches themselves. The aeration trenches are made use of only during the monsoon phase. After the break of the rains, new active roots are always found in or near the trenches first, after which a certain amount of development takes place under the grass.

The deep root system of the trees provided with aeration trenches behaved exactly like the controls.

THE RESULTS OBTAINED

The general results obtained with clean cultivation, grass, and grass with aeration trenches are shown in Plate IV, in which representative trees from the various plots have been drawn to scale. The drawings give a good idea of the main results of the experiment, namely: (1) the extremely deleterious effect of grass on young trees, (2) the less harmful effect of the same treatment on mature trees; (3) the partial recovery which sometimes takes place from the aeration trenches; and (4) the exceptional nature of the results with the guava, where the trees are able to grow under grass, but with reduced vigour, and where the aeration trenches have had little or no effect.

As would be expected from these results even a temporary removal of the grass cover has a profound effect. Whenever the roots of a tree under grass are exposed (for which purpose the grass has to be removed for a few days) there is an immediate increase in growth, accompanied by the formation of larger and darker-coloured leaves. The effect is clearly visible in the foliage above the excavation for as long as two years, but the rest of the tree is not affected.

THE CAUSE OF THE HARMFUL EFFECT OF GRASS

The examination of the root system of these eight species suggested that the first step in working out the cause of the harmful effect of grass would be to make a periodical examination of the soil gases. Determinations of the amount of CO_2 in the soil-air at a depth of 9 to 12 inches were carried out during 1919 under grass, under grass with aeration trenches, and under cultivated soil. About 10 litres of air were drawn out of the soil at each determination and passed through standard baryta which was afterwards titrated in the ordinary way. The 1919 results are given in Table 9 (p. 131) and are set out graphically in Fig. 5 (p. 132).

The results of 1920 and 1921 confirm these figures in all respects. Table 9 shows that during the monsoon the volume of carbon dioxide in the pore spaces under grass is increased about fivefold in comparison with the soil-air of cultivated land. As this gas is far more soluble in water than oxygen, the amounts of carbon dioxide actually dissolved in the water-films in which the root-hairs work would be much higher than the figures in the table suggest.

The production of large amounts of carbon dioxide in the soil-air during the rains would also affect the formation of humus, nitrification, and the mycorrhizal relationship, all of which depend on adequate aeration. Considerable progress was made in the investigation of the supply of combined nitrogen. At all periods of the year, except at the break of the rains, the amount of nitric nitrogen in the upper 18 inches of soil under grass varied from 10 to 20 per cent. of that met with in the cultivated plots. When the shortage of nitrogen in the case of the guava was made up by means of sulphate of ammonia during the rains of 1923, the trees

TABLE 9

Percentage by volume of carbon dioxide in the soil-gas, under grass and clean cultivation, Pusa, 1919[1]

Date and month when soil-gas was aspirated and analysed	Plot no. 1 grassed	Plot no. 2 grassed but partially aerated by trenches	Plot no. 3 surface cultivated	Rainfall in inches since January 1st, 1919
January 13, 14, and 17 .	0·444	0·312	0·269	Nil
February 20 and 21 .	0·472	0·320	0·253	1·30
March 21 and 22 . .	0·427	0·223	0·197	1·33
April 23 and 24 . .	0·454	0·262	0·203	2·69
May 16 and 17 . .	0·271	0·257	0·133	3·26
June 17 and 18 . .	0·341	0·274	0·249	4·53
July 17 and 18 . .	1·540	1·090	0·304	14·61
August 25 and 26 . .	1·590	0·836	0·401	23·29
September 19 and 20 .	1·908	0·931	0·450	30·67
October 21 and 22 .	1·297	0·602	0·365	32·90
November 14 and 15 .	0·853	0·456	0·261	32·90
December 22 and 23 .	0·398	0·327	0·219	32·92

under grass at once responded and produced fruit and foliage hardly distinguishable in size from the controls.

In the case of the litchi and loquat, the roots of which are unable to aerate themselves in the rains by forcing their way through the grass to the surface, heavy applications of combined nitrogen improved the growth, but a distinctly harmful effect remained— the manured trees as regards size and colour of the leaves, time of flowering, and production of new shoots occupying an intermediate position between the unmanured trees under grass and those under clean cultivation. These results are very similar to those obtained with apples at Cornell. At both places grass led to the disappearance of nitrates in the soil and restricted root development. The effect was only partially removed by the addition

[1] The determinations were carried out by Mr. Jatindra Nath Mukerjee of the Chemical Section, Pusa.

of nitrate of soda. In the guava, however, combined nitrogen removes the harmful effect because the roots of this tree are able to obtain all the oxygen they need. The guava, therefore, suffers from only one of the factors resulting from a grass carpet—lack of

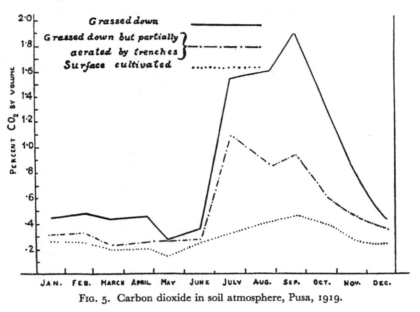

FIG. 5. Carbon dioxide in soil atmosphere, Pusa, 1919.

nitrate. The litchi and the loquat suffer from another factor as well—lack of oxygen.

FOREST TREES AND GRASS

Although the grass carpet acts as an asphyxiating agent to the roots of all the fruit-trees investigated except the guava, the ordinary Indian forest trees thrive under grass. Between the years 1921 and 1923 the relation between the grass carpet and the roots of the following fifteen forest trees was investigated (Table 10, p. 133). All thrive remarkably well under grass and show none of the harmful effects exhibited by fruit-trees.

Most of the forest trees in the plains of India flower and come into new leaf in the hot season and then proceed to form new shoots. After the early rains a distinct change is visible in the size, colour, and appearance of the foliage. The leaves become darker

PLATE VI

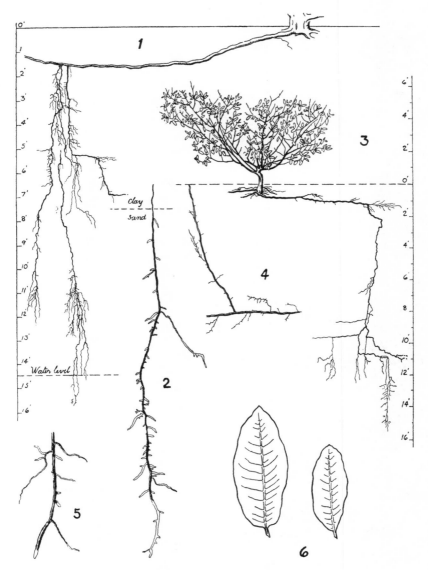

GUAVA (*Psidium Guyava*, L.)

FIG. 1. Superficial and deep roots (November 23, 1921).
FIG. 2. The influence of soil texture on the formation of the rootlets (March 29, 1921).
FIG. 3. The root-system under grass (April 21, 1921).
FIG. 4. Superficial rootlet growing to the surface (August 28, 1921).
FIG. 5. Formation of new rootlets in fine sand following the fall of the ground-water (November 20, 1921).
FIG. 6. Reduction in the size of leaves after 20 months under grass (right).

and more glossy; the story told by the young shoots of the custard apple (p. 122) is repeated.

Examination of the superficial root systems of the fifteen species during the rains of 1922 and 1923 yielded remarkably uniform results. All the trees produced abundant, normally developed

TABLE 10

Forest trees under grass in the Botanical Area, Pusa

Species	Time of flowering	Time of leaf-fall
Polyalthia longifolia Benth. & Hook, f. .	February–April	April
Melia Azadirachta L. . . .	March–May	March
Ficus bengalensis L.	April–May	March
Ficus religiosa L.	April–May	December
Ficus infectoria Roxb. . . .	February–May	December–January
Millingtonia hortensis Linn., f. . .	November–December	March
Butea frondosa Roxb. . . .	March	February
Phyllanthus Emblica L. . . .	March–May	February
Tamarindus indica L. . . .	April–June	March–April
Tectonia grandis Linn., f. . . .	July–August	February–March
Thespesia populnea Corr. . . .	Throughout the year but chiefly in the cold season	April
Pterospermum acerifolium Willd. . .	March–June	January–February
Wrightia tomentosa Roem. & Schult. .	April–May	January–February
Lagerstroemia Flos-Regina Retz . .	May	December–January
Dalbergia Sissoo Roxb. . . .	March	December–January

active rootlets in the upper 2 or 3 inches of soil and also on the surface; they therefore compete successfully with grass both for oxygen and nitrates. The large superficial roots were also well developed and compared favourably with the corresponding root system of fruit-trees under clean cultivation. The grass carpet had apparently no harmful effect on the root system near the surface.

Between the hot weather of 1921 and the early months of 1924 the complete root systems of these fifteen species were investigated. In all cases the large surface roots gave off thin branches which grew vertically downwards to the cold-season level of the ground water. Root activity in all cases was practically confined in the hot season to the deep moist layers of sand between 10 and 20 feet below the surface, the roots always making the fullest use of the tunnels of *Termites* and other burrowing insects for passing easily through clay layers from one zone of sandy soil to the next below.

Cavities in the soil were always fully used for root development. Soon after the rains the dormant surface roots burst into activity. As the ground water rose the deep root system became dormant; in August the active surface roots always showed marked aerotropism. The formation of nitrates which takes place about the time the cold-season crops are sown was followed by a definite burst of renewed root activity in the surface soil, followed by the production of new shoots and leaves. As the ground water falls in the autumn and the soil draws in oxygen, the formation of active roots follows the descending water-table exactly as has been described in the case of the guava.

The facts of root distribution and periodicity in root activity in forest trees explain why these trees do so well under grass and are able to vanquish it if allowed free competition. The chief weapons which enable forest trees to oust grasses and herbs from the habitat are the following:

1. The deep root system admits of growth during the dry season when the grass is dormant, thereby enabling the trees to utilize moisture and food materials in the soil down to at least 20 feet. This markedly extends the period of assimilation.

2. The habit of trees is a great advantage in the struggle for light.

3. The active roots of the surface system are resistant to poor soil aeration, and are able to reach the surface and compete successfully with the grass for oxygen and for minerals.

The character which distinguishes forest trees from fruit-trees is the power possessed by the surface roots of the former to avoid the consequences of poor soil aeration by forcing their way through a grass carpet in active growth to the air and to obtain oxygen as well as a share of the nitrates in the surface soil. The surface roots of most fruit-trees are very susceptible to carbon dioxide and try to avoid it by growing downwards. The trees are therefore deprived of oxygen and of combined nitrogen during the rains, and slowly starve. The guava is an exception among fruit-trees. Here the active roots reach the surface in the rains and the trees are able to maintain themselves. This explains why the pastures of Grenada and St. Vincent in the West Indies are so rapidly invaded and destroyed by the wild guava. The hedgerows and pastures of Great Britain if left to themselves behave in a similar way. The hedgerows soon begin to invade the fields. Young trees make their appearance; grass areas become woodland. The transformation, however, is much slower in Great Britain than in the tropics.

These studies on the root development of tropical forest trees throw a good deal of light on the soil aeration factor and the part the plant can play in such investigations. The movement of the ground water affects soil aeration directly. The two periods—the beginning and end of the rainy season—when the surface soil contains abundant air and ample moisture and when the temperature is favourable for nitrification, correspond exactly with times when nitrates accumulate (p. 209) and when growth is at its maximum. When soil aeration is interfered with during the rains by two factors, (1) the rise of the ground water, and (2) the formation of colloids in the surface soil, the plant roots respond by growing to the surface. Root development, therefore, is an important instrument in such an investigation when examined throughout the year.

The root development of trees influences the maintenance of soil fertility in the plains of India and indeed in many other regions. The dead roots provide the deeper layers of soil with organic matter and an almost perfect drainage and aerating system. The living roots comb the upper 20 feet of soil for such minerals as phosphates and potash which are used in the green leaves. These leaves in due course are converted into humus and help to enrich the surface soil. This explains why the soils of North Bihar, although very low in total and available phosphates, are so exceedingly fertile and yield heavy crops without any addition of mineral manures. The figures given by the analysis of the surface soil must be repeated in the lower layers and should be interpreted not in terms of the upper 9 inches but of the upper 20 feet.

The tree is the most efficient agent available for making use of the minerals in the soil. It can grow almost anywhere, it will vanquish most of the other forms of vegetation, and it will leave the soil in a highly fertile condition. It follows therefore that the trees and shrubs of the hedgerows, parks, and woodlands of countries like Great Britain must continue to be used for the maintenance of soil fertility. In Saxon times most of our best land was under forest. The fertility stored in the soil made the gradual clearing of this woodland worth while. In the future, when agriculture comes into its own and when it is no longer regarded solely as an industry, it may be desirable to embark on long term rotations in which woods and park-land are turned into arable, and worn-out arable back into woodland or into mixed grass and trees. In this way the root system of the tree can be used to restore soil fertility.

THE AERATION OF THE SUB-SOIL

One of the universal methods of improving aeration is sub-soiling. The methods adopted vary greatly according to the factor which has interfered with aeration and the means available for improving the air-supply.

In temperate regions the chief factor which cuts off the sub-soil from the atmosphere is shortage of humus aggravated by impermeable pans (produced by the plough and by the soil particles themselves) or a permanent grass carpet accompanied by the constant treading of animals. The result in all cases is the same—the supply of air to the sub-soil is reduced.

In loamy soils plough-pans develop very rapidly if the content of organic matter falls off and the earthworm population declines. A well-defined zone of close and sticky soil is formed just under the plough sole which holds up water, thereby partly asphyxiating the sub-soil below and water-logging the soil above.

In sandy soils as well as in silts, pans are formed with the greatest ease from the running together of the particles, particularly when artificials take the place of farm-yard manure and the temporary ley is not properly utilized. One of the most interesting cases of pan formation that I have observed in Great Britain was on the permanent manurial plots of the Woburn Experiment Station, where an attempt to grow cereals year after year on the greensand by means of artificial manures has been followed by complete failure of the crop. The soil has gone on permanent strike. The destruction of the earthworm population by the regular application of chemicals had deprived the land of its natural aerating agencies. Failure to renew the organic matter by a suitable rotation had resulted in a soil devoid of even a trace of tilth. About 9 inches below the surface, a definite pan (made up of sand particles loosely cemented together) occurred, which had so altered the aeration of the sub-soil that the whole of these experimental plots were covered with a dense growth of mares' tail (*Equisetum arvense* L.), a perennial weed which always indicates a badly aerated sub-soil. Nature as usual had summed up the position in her own inimitable fashion. There was no need of tabulated yields, analyses, curves, and statistics to explain the consequences of improper methods of agriculture.

The conventional method of dealing with arable pans in this country is by means of some sub-soiling implement which breaks them up and restores aeration. This should be accompanied whenever possible by heavy dressings of farm-yard manure, so that the tilth can be improved and the earthworm population restored. Some deep-rooted crop like lucerne, or even a temporary ley, should be called in to complete the cure. Sub-soiling heavy land under grass is proving even more advantageous than on arable areas. This leads, as we have seen (p. 102), to humus formation under the turf and to an increase in the stock-carrying capacity of the land.

In the East the ventilation of the sub-soil is perhaps even more important than in the West. In India, for example, one of the common consequences of the monsoon rainfall and of flooding the surface with irrigation water is pan formation on a colossal scale due to the formation of soil colloids—the whole of the surface soil tends to become a pan. This has to be broken up. The cultivators of the Orient set about this task in a very interesting way. Whenever they can use the roots of a leguminous crop as a sub-soiler they invariably employ this machine. It has the merit of costing nothing, of yielding essential food and fodder, and of suiting the small field. In the Indo-Gangetic plain the universal sub-soiler is the pigeon pea, the roots of which not only break up soil pans with ease but also add organic matter at the same time. On the Western frontier the sub-soiling of the dense loess soils is always done by the roots of a lucerne crop. On the black cotton soils of Peninsular India where the monsoon rainfall converts the whole of the surface soil into a vast colloidal pan, the agricultural situation is saved by the succeeding hot season which dries out this pan and reduces its volume to such an extent that a multitude of deep fissures occur right down to the sub-soil. The black soils of India plough and sub-soil themselves. The moist winds, which precede the south-west monsoon in May and early June, replace some of the lost moisture; the heavy clods break down and when the early rains arrive a magnificent tilth can be prepared for the cotton crop. The sub-soiling in this case is done by Nature; the cultivators merely give a subsequent cultivation and then sow the crop.

BIBLIOGRAPHY

CLEMENTS, F. E. *Aeration and Air Content: the Role of Oxygen in Root Activity*, Publication No. 315, Carnegie Institution of Washington, 1921.

HOWARD, A. *Crop Production in India: A Critical Survey of its Problems*, Oxford University Press, 1924.

—— 'The Effect of Grass on Trees', *Proc. Royal Soc.*, Series B, xcvii, 1925, p. 284.

LYON, T. L., HEINICKE, A. J., and WILSON, D. D. *The Relation of Soil Moisture and Nitrates to the Effects of Sod on Apple Trees*, Memoir 63, Cornell Agricultural Expt. Station, 1923.

THE DUKE OF BEDFORD, and PICKERING, S. U. *Science and Fruit-Growing*, London, 1919.

WEAVER, J. E., JEAN, F. C., and CRIST, J. W. *Development and Activities of Crop Plants*, Publication No. 316, Carnegie Institution of Washington, 1922.

CHAPTER X

SOME DISEASES OF THE SOIL
SOIL EROSION

PERHAPS the most widespread and the most important disease of the soil at the present time is soil erosion, a phase of infertility to which great attention is now being paid.

Soil erosion in the very mild form of denudation has been in operation since the beginning of time. It is one of the normal operations of Nature going on everywhere. The minute soil particles which result from the decay of rocks find their way sooner or later to the ocean, but many may linger on the way, often for centuries, in the form of one of the constituents of fertile fields. This phenomenon can be observed in any river valley. The fringes of the catchment area are frequently uncultivated hills through the thin soils of which the underlying rocks protrude. These are constantly weathered and in the process yield a continuous supply of minute fragments in all stages of decomposition.

The slow rotting of exposed rock surfaces is only one of the forms of decay. The covering of soil is no protection to the under-lying strata but rather the reverse, because the soil water, containing carbon dioxide in solution is constantly disintegrating the parent rock, first producing sub-soil and then actual soil. At the same time the remains of plants and animals are converted into humus. The fine soil particles of mineral origin, often mixed with fragments of humus, are then gradually removed by rain, wind, snow, or ice to lower regions. Ultimately the rich valley lands are reached where the accumulations may be many feet in thickness. One of the main duties of the streams and rivers, which drain the valley, is to transport these soil particles into the sea where fresh land can be laid down. The process looked at as a whole is nothing more than Nature's method of the rotation, not of the crop, but of the soil itself. When the time comes for the new land to be enclosed and brought into cultivation agriculture is born again. Such operations are well seen in England in Holbeach marsh and similar areas round the Wash. From the time of the Romans to the present day, new areas of fertile soil, which now fetch £100 an acre or even more, have been re-created from the uplands by the Welland, the Nen, and the Ouse. All this fertile land, perhaps the most valuable in England, is the result of two of the most widespread processes in Nature—weathering and denudation.

It is when the tempo of denudation is vastly accelerated by human agencies that a perfectly harmless natural process becomes transformed into a definite disease of the soil. The condition known as soil erosion—a man-made disease—is then established. *It is, however, always preceded by infertility*: the inefficient, overworked, dying soil is at once removed by the operations of Nature and hustled towards the ocean, so that new land can be created and the rugged individualists—the bandits of agriculture—whose cursed thirst for profit is at the root of the mischief can be given a second chance. Nature is anxious to make a new and better start and naturally has no patience with the inefficient. Perhaps when the time comes for a new essay in farming, mankind will have learnt a great lesson—how to subordinate the profit motive to the sacred duty of handing over unimpaired to the next generation the heritage of a fertile soil. Soil erosion is nothing less than the outward and visible sign of the complete failure of a policy. The causes of this failure are to be found in ourselves.

The damage already done by soil erosion all over the world looked at in the mass is very great and is rapidly increasing. The regional contributions to this destruction, however, vary widely. In some areas like north-western Europe, where most of the agricultural land is under a permanent or temporary cover crop (in the shape of grass or leys), and there is still a large area of woodland and forest, soil erosion is a minor factor in agriculture. In other regions like parts of North America, Africa, Australia, and the countries bordering the Mediterranean, where extensive deforestation has been practised and where almost uninterrupted cultivation has been the rule, large tracts of land once fertile have been almost completely destroyed.

The United States of America is perhaps the only country where anything in the nature of an accurate estimate of the damage done by erosion has been made. Theodore Roosevelt first warned the country as to its national importance. Then came the Great War with its high prices, which encouraged the wasteful exploitation of soil fertility on an unprecedented scale. A period of financial depression, a series of droughts and dust-storms, emphasized the urgency of the salvage of agriculture. During Franklin Roosevelt's Presidency, soil conservation has become a political and social problem of the first importance. In 1937 the condition and needs of the agricultural land of the U.S.A. were appraised. No less than 253,000,000 acres, or 61 per cent. of the total area under crops, had either been completely or partly destroyed or had lost most of its fertility. Only 161,000,000 acres, or 39 per cent. of the cultivated area, could be safely farmed by present methods. In less than a century the United States has therefore lost nearly three-fifths of its agricultural capital. If the whole of the potential resources of the country could be utilized and the best possible practices introduced everywhere, about 447,466,000 acres could be brought into use—an area somewhat greater than the present crop land area of 415,334,931 acres. The position therefore is not hopeless. It will, however, be very difficult, very expensive, and very time-consuming to restore the vast areas of eroded land even if money is no object and large amounts of manure are used and green-manure crops are ploughed under.

The root of this soil erosion trouble in the United States is mis-use of the land. The causes of this misuse include lack of individual knowledge of soil fertility on the part of the pioneers and their descendants; the traditional attitude which regarded the land as a source of profit; defects in farming systems, in tenancy, and finance—most mortgages contain no provisions for the main-tenance of fertility; instability of agricultural production (as carried out by millions of individuals), prices and income in con-trast to industrial production carried on by a few large corporations. The need for maintaining a correct relation between industrial and agricultural production so that both can develop in full swing on the basis of abundance has only recently been understood. The country was so vast, its agricultural resources were so immense, that the profit seekers could operate undisturbed until soil fertility —the country's capital—began to vanish at an alarming rate. The present position, although disquieting, is not impossible. The resources of the Government are being called up to put the land in order. The magnitude of the effort, the mobilization of all available knowledge, the practical steps that are being taken to save what is left of the soil of the country and to help Nature to repair the damage already done are graphically set out in *Soils and Men*, the Year Book of the United States Department of Agriculture of 1938. This is perhaps the best local account of soil erosion which has yet appeared.

The rapid agricultural development of Africa was soon followed by soil erosion. In South Africa, a pastoral country, some of the best grazing areas are already semi-desert. The Orange Free State in 1879 was covered with rich grass, interspersed with reedy pools, where now only useless gullies are found. Towards the end of the nineteenth century it began to be realized all over South Africa that serious over-stocking was taking place. In 1928 the Drought Investigation Commission reported that soil erosion was extending rapidly over many parts of the Union, and that the eroded material was silting up reservoirs and rivers and causing a marked decrease in the underground water-supplies. The cause of erosion was considered to be the reduction of vegetal cover brought about by incorrect veld management—the concentration of stock in kraals, over-stocking, and indiscriminate burning to obtain fresh autumn or winter grazing. In Basutoland, a normally well-watered country, soil erosion is now the most immediately pressing administrative problem. The pressure of population has brought large areas under the plough and has intensified over-stocking on the remaining pasture. In Kenya the soil erosion problem has become serious during the last three years, both in the native reserves and in the European areas. In the former, wealth depends on the possession of large flocks and herds; barter is carried on in terms of live stock; the bride price is almost universally paid in animals; numbers rather than quality are the rule. The natural consequence is over-stocking, over-grazing, and the destruction of the natural covering of the soil. Soil erosion is the inevitable result. In the European areas erosion is caused by long and continuous over-cropping without the adoption of measures to prevent the loss of soil and to maintain the humus content. Locusts have of late been responsible for greatly accelerated erosion; examples are to be seen where the combined effect of locusts and goats has resulted in the loss of a foot of surface soil in a single rainy season.

The countries bordering the Mediterranean provide striking examples of soil erosion, accompanied by the formation of deserts which are considered to be due to one main cause—the slow and continuous deforestation of the last 3,000 years. Originally well wooded, no forests are to be found in the Mediterranean region proper. Most of the original soil has been washed away by the sudden winter torrents. In North Africa the fertile cornfields, which existed in Roman times, are now desert. Ferrari in his book on woods and pastures refers to the changes in the soil and climate of Persia after its numerous and majestic parks were destroyed; the soil was transformed into sand; the climate became arid and suffocating; springs first decreased and then disappeared. Similar changes took place in Egypt when the forests were devastated; a decrease in rainfall and in soil fertility was accompanied by loss of uniformity in the climate. Palestine was once covered with valuable forests and fertile pastures and possessed a cool and moderate climate; to-day its mountains are denuded, its rivers are almost dry, and crop production is reduced to a minimum.

The above examples indicate the wide extent of soil erosion, the very serious damage that is being done, and the fundamental cause of the trouble—misuse of the land. In dealing with the remedies which have been suggested and which are now being tried out, it is essential to envisage the real nature of the problem. It is nothing less than the repair of Nature's drainage system—the river—and of Nature's method of providing the country-side with a regular water-supply. The catchment area of the river is the natural unit in erosion control. In devising this control we must restore the efficiency of the catchment area as a drain and also as a natural storage of water. Once this is accomplished we shall hear very little about soil erosion.

Japan provides perhaps the best example of the control of soil erosion in a country with torrential rains, highly erodible soils, and a topography which renders the retention of the soil on steep slopes very difficult. Here erosion has been effectively held in check, by methods adopted regardless of cost, for the reason that the alternative to their execution would be national disaster. The great danger from soil erosion in Japan is the deposition of soil debris from the steep mountain slopes on the rice-fields below. The texture of the rice soils must be maintained so that the fields will hold water and allow of the minimum of through drainage. If such areas became covered with a deep layer of permeable soil, brought down by erosion from the hill-sides, they would no longer hold water, and rice cultivation—the mainstay of Japan's food-supply—would be out of the question. For this reason the country has spent as much as ten times the capital value of eroding land on soil conservation work, mainly as an insurance for saving the valuable rice lands below. Thus in 1925 the Tokyo Forestry Board spent 453 yen (£45) per acre in anti-erosion measures on a forest area, valued at 40 yen per acre, in order to save rice-fields lower down valued at 240 to 300 yen per acre.

The dangers from erosion have been recognized in Japan for centuries and an exemplary technique has been developed for preventing them. It is now a definite part of national policy to maintain the upper regions of each catchment area under forest, as the most economical and effective method of controlling flood waters and insuring the production of rice in the valleys. For many years erosion control measures have formed an important item in the national budget.

According to Lowdermilk, erosion control in Japan is like a game of chess. The forest engineer, after studying his eroding valley, makes his first move, locating and building one or more check dams. He waits to see what Nature's response is. This determines the forest engineer's next move, which may be another dam or two, an increase in the former dam, or the construction of side retaining walls. After another pause for observation, the next move is made and so on until erosion is checkmated. The operation of natural forces, such as sedimentation and re-vegetation, are guided and used to the best advantage to keep down costs and to obtain practical results. *No more is attempted than Nature has already done in the region.* By 1929 nearly 2,000,000 hectares of protection forests were used in erosion control. These forest areas do more than control erosion. They help the soil to absorb and maintain large volumes of rain-water and to release it slowly to the rivers and springs.

China, on the other hand, presents a very striking example of the evils which result from the inability of the administration to deal with the whole of a great drainage unit. On the slopes of the upper reaches of the Yellow River extensive soil erosion is constantly going on. Every year the river transports over 2,000 million tons of soil, sufficient to raise an area of 400 square miles by 5 feet. This is provided by the easily erodible loess soils of the upper reaches of the catchment area. The mud is deposited in the river bed lower down so that the embankments which contain the stream have constantly to be raised. Periodically the great river wins in this unequal contest and destructive inundations result. The labour expended on the embankments is lost because the nature of the erosion problem as a whole has not been grasped, and the area drained by the Yellow River has not been studied and dealt with as a single organism. The difficulty now is the over-population of the upper reaches of the catchment area, which prevents afforestation and laying down to grass. Had the Chinese maintained effective control of the upper reaches—the real cause of the trouble—the erosion problem in all probability would have been solved long ago at a lesser cost in labour than that which has been devoted to the embankment of the river. China, unfortunately, does not stand alone in this matter. A number of other rivers, like the Mississippi, are suffering from overwork, followed by periodical floods as the result of the growth of soil erosion in the upper reaches.

Although the damage done by uncontrolled erosion all over the world is very great, and the case for action needs no argument, nevertheless there is one factor on the credit side which has been overlooked in the recent literature. A considerable amount of new soil is being constantly produced by natural weathering agencies from the sub-soil and the parent rock. This when suitably conserved will soon re-create large stretches of valuable land. One of the best regions for the study of this question is the black cotton soil of Central India, which overlies the basalt. Here, although erosion is continuous, the soil does not often disappear altogether, for the reason that as the upper layers are removed by rain, fresh soil is re-formed from below. The large amount of earth so produced is well seen in the Gwalior State, where the late Ruler employed an irrigation officer, lent by the Government of India, to construct a number of embankments, each furnished with spill-ways, across many of the valleys, which had suffered so badly by uncontrolled rain-wash in the past that they appeared to have no soil at all, the scrub vegetation just managing to survive in the crevices of the bare rock. How great is the annual formation of new soil, even in such unpromising circumstances, must be seen to be believed. In a very few years, the construction of embankments was followed by stretches of fertile land which soon carried fine crops of wheat. A brief illustrated account of the work done by the late Maharaja of Gwalior would be of great value at the moment for introducing a much needed note of optimism in the consideration of this soil erosion problem. Things are not quite so hopeless as they are often made to appear.

Why is the forest such an effective agent in the prevention of soil erosion and in feeding the springs and rivers? The forest does two things: (1) the trees and undergrowth break up the rain-fall into fine spray and the litter on the ground protects the soil from erosion; (2) the residues of the trees and animal life met with in all woodlands are converted into humus, which is then absorbed by the soil underneath, increasing its porosity and water-holding power. The soil cover and the soil humus together prevent erosion and at the same time store large volumes of water. These factors—soil protection, soil porosity, and water retention—conferred by the living forest cover, provide the key to the solution of the soil erosion problem. All other purely mechanical remedies such as terracing and drainage are secondary matters, although of course important in their proper place. The soil must have as much cover as possible; it must be well stocked with humus so that it can drink in and retain the rainfall. It follows, therefore, that in the absence of trees there must be a grass cover, some cover-crop, and ample provision for keeping up the supply of humus. Each field so provided suffers little or no erosion. This confirms the view of Williams (Timiriasev Academy, Moscow) who, before erosion became important in the Soviet Union, advanced an hypothesis that the decay of past civilizations was due to a decline in soil fertility, consequent on the destruction of the soil's crumb structure when the increasing demands of civilization necessitated the wholesale ploughing up of grass-land. Williams regarded grass as the basis of all agricultural land utilization and the soil's chief weapon against the plundering instincts of humanity. His views are exerting a marked influence on soil conservation policy in the U.S.S.R. and indeed apply to many other countries.

Grass is a valuable factor in the correct design and construction of surface drains. Whenever possible these should be wide, very shallow, and completely grassed over. The run-off then drains away as a thin sheet of clear water, leaving all the soil particles behind. The grass is thereby automatically manured and yields abundant fodder. This simple device was put into practice at the Shahjahanpur Sugar Experiment Station in India. The earth service roads and paths were excavated so that the level was a few inches below that of the cultivated area. They were then grassed over, becoming very effective drains in the rainy season, carrying off the excess rainfall as clear water without any loss of soil.

If we regard erosion as the natural consequence of improper methods of agriculture, and the catchment area of the river as the natural unit for the application of soil conservation methods, the various remedies available fall into their proper place. The upper reaches of each river system must be afforested; cover crops including grass and leys must be used to protect the arable surface whenever possible; the humus content of the soil must be increased and the crumb structure restored so that each field can drink in its own rainfall; over-stocking and over-grazing must be prevented; simple mechanical methods for conserving the soil and regulating the run-off, like terracing, contour cultivation and contour drains, must be utilized. There is, of course, no single anti-erosion device which can be universally adopted. The problem must, in the nature of things, be a local one. Nevertheless, certain guiding principles exist which apply everywhere. First and foremost is the restoration and maintenance of soil fertility, so that each acre of the catchment area can do its duty by absorbing its share of the rainfall.

THE FORMATION OF ALKALI LANDS

When the land is continuously deprived of oxygen the plant is soon unable to make use of it: a condition of permanent infertility results.

In many parts of the tropics and sub-tropics agriculture is interfered with by accumulations of soluble salts composed of various mixtures of the sulphate, chloride, and carbonate of sodium. Such areas are known as alkali lands. When the alkali phase is still in the mild or incipient stage, crop production becomes difficult and care has to be taken to prevent matters from getting worse. When the condition is fully established, the soil dies; crop production is then out of the question. Alkali lands are common in Central Asia, India, Persia, Iraq, Egypt, North Africa, and the United States.

At one period it was supposed that alkali soils were the natural consequences of a light rainfall, insufficient to wash out of the land the salts which always form in it by progressive weathering of the rock powder of which all soils largely consist. Hence alkali lands were considered to be a natural feature of arid tracts, such as parts of north-west India, Iraq, and northern Africa, where the rainfall is very small. Such ideas on the origin and occurrence of alkali lands do not correspond with the facts and are quite misleading. The rainfall of the Province of Oudh, in India, for example, where large stretches of alkali lands naturally occur, is certainly adequate to dissolve the comparatively small quantities of soluble salts found in these infertile areas, if their removal were a question of sufficient water only. In North Bihar the average rainfall, in the sub-montane tracts where large alkali patches are common, is about 50 to 60 inches a year. Arid conditions, therefore, are not essential for the production of alkali soils; heavy rainfall does not always remove them. What is a necessary condition is impermeability. In India whenever the land loses its porosity, by the constant surface irrigation of stiff soils with a tendency to impermeability, by the accumulation of stagnant subsoil water, or through some interference with the surface drainage, alkali salts sooner or later appear. Almost any agency, even over-cultivation and over-stimulation by means of artificial manures, both of which oxidize the organic matter and slowly destroy the crumb structure, will produce alkali land. In the neighbourhood of Pusa in North Bihar, old roads and the sites of bamboo clumps and of certain trees such as the tamarind (*Tamarindus indica* L.) and the *pipul* (*Ficus religiosa* L.), always give rise to alkali patches when they are brought into cultivation. The densely packed soil of such areas invariably shows the bluish-green markings which are associated with the activities of those soil organisms which live in badly aerated soils without a supply of free oxygen. A few inches below the alkali patches, which occur on the stiff loess soils of the Quetta Valley, similar bluish-green and brown markings always occur. In the alkali zone in North Bihar, wells have always to be left open to the air, otherwise the water is contaminated by sulphuretted hydrogen, thereby indicating a well-marked reductive phase in the deeper layers. In a sub-soil drainage experiment on the black soils of the Nira valley in Bombay where perennial irrigation was followed by the formation of alkali land, Mann and Tamhane found that the salt water which ran out of these drains soon smelt strongly of sulphuretted hydrogen, and a white deposit of sulphur was formed at the mouth of each drain, proving how strong were the reducing actions in this soil. Here the reductive phase in alkali formation was unconsciously demonstrated in an area where alkali salts were unknown until the land was water-logged by over-irrigation and the oxygen-supply of the soil was restricted.

The view that the origin of alkali land is bound up with defective soil aeration is supported by the recent work on the origin of salt-water lakes in Siberia. In Lake Szira-Kul, between Bateni and the mountain range of Kizill Kaya, Ossendowski observed in the black ooze taken from the bottom of the lake and in the water a certain distance from the surface an immense network of colonies of sulphur bacilli which gave off large quantities of sulphuretted hydrogen and so destroyed practically all the fish in this lake. The great water basins in Central Asia are being metamorphosed in a similar way into useless reservoirs of salt water, smelling strongly of hydrogen sulphide. In the limans near Odessa and in portions of the Black Sea, a similar process is taking place. The fish, sensing the change, are slowly leaving this sea as the layers of water, poisoned by sulphuretted hydrogen, are gradually rising towards the surface. The death of the lakes scattered over the immense plains of Asia and the destruction of the impermeable soils of this continent from alkali salt formation are both due to the same primary cause—intense oxygen starvation. Often this oxygen starvation occurs naturally; in other cases it follows perennial irrigation.

The stages in the development of the alkali condition are somewhat as follows. The first condition is an impermeable soil. Such soils—the *usar* plains of northern India for example—occur naturally where the climatic conditions favour those biological and physical factors which destroy the soil structure by disintegrating the compound particles into their ultimate units. These latter are so extremely minute and so uniform in size that they form with water a mixture possessing some of the properties of colloids which, when dry, pack into a hard dry mass, practically impermeable to water and very difficult to break up. Such soils are very old. They have always been impermeable and have never come into cultivation.

In addition to the alkali tracts which occur naturally a number are in course of formation as the result of errors in soil management, the chief of which are:

(*a*) The excessive use of irrigation water. This gradually destroys the binding power of the organic cementing matter which glues the soil particles together, and displaces the soil air. Anaerobic changes, indicated by blue and brownish markings, first occur in the lower layers and finally lead to the death of the soil. It is this slow destruction of the living soil that must be prevented if the existing schemes of perennial irrigation are to survive. The process is taking place before our eyes to-day in the Canal Colonies of India where irrigation is loosely controlled.

(*b*) Over-cultivation without due attention to the replenishment of humus. In those continental areas like the Indo-Gangetic plain, where the risk of alkali is greatest, the normal soils contain only a small reserve of humus, because the biological processes which consume organic matter are very intense at certain seasons due to sudden changes from low to very high temperatures and from intensely dry weather to periods of moist tropical conditions. Accumulations of organic matter such as occur in temperate zones are impossible. There is, therefore, a very small margin of safety. The slightest errors in soil management will not only destroy the small reserve of humus in the soil but also the organic cement on which the compound soil particles and the crumb structure depend.

The result is impermeability, the first stage in the formation of alkali salts.

(c) The use of artificial manures, particularly sulphate of ammonia. The presence of additional combined nitrogen in an easily assimilable form stimulates the growth of fungi and other organisms which, in the search for the organic matter needed for energy and for building up microbial tissue, use up first the reserve of soil humus and then the more resistant organic matter which cements the soil particles. Ordinarily this glue is not affected by the processes going on in a normally cultivated soil, but it cannot withstand the same processes when stimulated by dressings of artificial manures.

Alkali land therefore starts with a soil in which the oxygen-supply is permanently cut off. Matters then go from bad to worse very rapidly. All the oxidation factors which are essential for maintaining a healthy soil cease. A new soil flora—composed of anaerobic organisms which obtain their oxygen from the sub-stratum—is established. A reduction phase ensues. The easiest source of oxygen—the nitrates—is soon exhausted. The organic matter then undergoes anaerobic fermentation. Sulphuretted hydrogen is produced as the soil dies, just as in the lakes of Central Asia. The final result of the chemical changes that take place is the accumulation of the soluble salts of alkali land—the sulphate, chloride, and carbonate of sodium. When these salts are present in injurious amounts they appear on the surface in the form of snow-white and brownish-black incrustations. The former (white alkali) consists largely of the sulphate and chloride of sodium, and the latter (the dreaded black alkali) contains sodium carbonate in addition and owes its dark colour to the fact that this salt is able to dissolve the organic matter in the soil and produce physical conditions which render drainage impossible. According to Hilgard, sodium carbonate is formed from the sulphate and chloride in the presence of carbon dioxide and water. The action is reversed in the presence of oxygen. Subsequent investigations have modified this view and have shown that the formation of sodium carbonate in soil takes place in stages. The appearance of this salt always marks the end of the chapter. The soil is dead. Reclamation then becomes difficult on account of the physical conditions set up by these alkali salts and the dissolved organic matter.

The occurrence of alkali land, as would be expected from its origin, is extremely irregular. When ordinary alluvial soils like those of the Punjab and Sind are brought under perennial irrigation, small patches of alkali first appear where the soil is heavy; on stiffer areas the patches are large and tend to run together. On open permeable stretches, on the other hand, there is no alkali. In tracts like the Western Districts of the United Provinces, where irrigation has been the rule for a long period, zones of well-aerated land carrying fine irrigated crops occur alongside the barren alkali tracts. Iraq also furnishes interesting examples of the connexion between alkali and poor soil aeration. Intensive cultivation under irrigation is only met with in that country where the soils are permeable and the natural drainage is good. Where the drainage and aeration are poor, the alkali condition at once becomes acute. There are, of course, a number of irrigation schemes, such as the staircase cultivation of the Hunzas in north-west India and of Peru, where the land has been continually watered from time immemorial without any development of alkali salts. In Italy and Switzerland perennial irrigation has been practised for long periods without harm to the soil. In all such cases, however, careful attention has been paid to drainage and aeration and to the maintenance of humus; the soil processes have been confined by Nature or by man to the oxidative phase; the cement of the compound particles has been protected by keeping up a sufficiency of organic matter.

Every possible gradation in alkali land is met with. Minute quantities of alkali salts in the soil have no injurious effect on crops or on the soil organisms. It is only when the proportion increases beyond a certain limit that they first interfere with growth and finally prevent it altogether. Leguminous crops are particularly sensitive to alkali especially when this contains carbonate of soda. The action of alkali salts on the plant is a physical one and depends on the osmotic pressure of solutions, which increases with the amount of the dissolved substance. For water to pass readily from the soil into the roots of plants, the osmotic pressure of the cells of the root must be considerably greater than that of the soil solution outside. If the soil solution became stronger than that of the cells, water would pass backwards from the roots to the soil and the crops would dry up. This state of affairs naturally occurs when the soil becomes charged with alkali salts beyond a certain point. The crops are then unable to take up water and death results. The roots behave like a plump strawberry when placed in a strong solution of sugar. Like the strawberry they shrink in size because they have lost water to the stronger solution outside. Too much salt in the water therefore makes irrigation water useless and destroys the canal as a commercial proposition.

The reaction of the crop to the first stages in alkali production is interesting. For twenty years at Pusa and eight years in the Quetta Valley I had to farm land, some of which hovered, as it were, on the verge of alkali. The first indication of the condition is a darkening of the foilage and the slowing down of growth. Attention to soil aeration, to the supply of organic matter, and to the use of deep-rooting crops like lucerne and pigeon pea, which break up the sub-soil, soon sets matters right. Disregard of Nature's danger signals, however, leads to trouble—a definite alkali patch is formed. When cotton is grown under canal irrigation on the alluvial soils of the Punjab, the reaction of the plant to incipient alkali is first shown by the failure to set seed, on account of the fact that the anther, the most sensitive portion of the flower, fails to function and to liberate its pollen. The cotton plant naturally finds it difficult to obtain from mild alkali soil all the water it needs—this shortage is instantly reflected in the breakdown of the floral mechanism.

The theory of the reclamation of alkali land is very simple. All that is needed, after treating the soil with sufficient gypsum (which transforms the sodium clays into calcium clays), is to wash out the soluble salts, to add organic matter, and then to farm the land properly. Such reclaimed soils are then exceedingly fertile and remain so. If sufficient water is available it is sometimes possible to reclaim alkali soils by washing only. I once confirmed this. The berm of a raised water channel at the Quetta Experiment Station was faced with rather heavy soil from an alkali patch. The constant passage of the irrigation water down the water channel soon removed the alkali salts. This soil then produced some of the heaviest crops of grass I have ever seen in the tropics. When, however, the attempt is made to reclaim alkali areas on a field scale, by flooding and draining, difficulties at once arise unless steps are taken first to replace all the sodium in the soil complex by calcium and then to prevent the further formation of sodium clays. Even when these reclamation methods succeed, the cost is always considerable; it soon becomes prohibitive; the game is not worth the candle. The removal of the alkali salts is only the first step; large quantities of organic matter are then needed; adequate soil aeration must be provided; the greatest care must be taken to preserve these reclaimed soils and to see that no reversion to the alkali condition occurs. It is exceedingly easy under canal irrigation to create alkali salts on certain areas. It is exceedingly difficult to reverse the process and to transform alkali land back again into a fertile soil.

Nature has provided, in the shape of alkali salts, a very effective censorship for all schemes of perennial irrigation. The conquest of the desert, by means of the canal, by no means depends on the mere provision of water and arrangements for the periodical flooding of the surface. This is only one of the factors of the problem. The water must be used in such a manner and the soil management must be such that the fertility of the soil is maintained intact. There is obviously no point in creating, at vast expense, a Canal Colony and producing crops for a generation or two, followed by a desert of alkali land. Such an achievement merely provides another example of agricultural banditry. It must always be remembered that the ancient irrigators never developed any efficient method of perennial irrigation, but were content with the basin system,[1] a device by which irrigation and soil aeration can be combined. In his studies on irrigation and drainage, King concludes an interesting discussion of this question in the following words, which deserve the fullest consideration on the part of the irrigation authorities all over the world:

'It is a noteworthy fact that the excessive development of alkalis in India, as well as in Egypt and California, is the result of irrigation

[1] The land is embanked; watered once; when dry enough it is cultivated and sown. In this way water can be provided without any interference with soil aeration.

practices modern in their origin and modes and instituted by people lacking in the traditions of the ancient irrigators, who had worked these same lands thousands of years before. The alkali lands of to-day, in their intense form, are of modern origin, due to practices which are evidently inadmissible, and which in all probability were known to be so by the people whom our modern civilization has supplanted.'

BIBLIOGRAPHY

SOIL EROSION

GORRIE, R. M. 'The Problem of Soil Erosion in the British Empire, with special reference to India', *Journal of the Royal Society of Arts*, lxxxvi, 1938, p. 901.

HOWARD, SIR ALBERT. 'A Note on the Problem of Soil Erosion', *Journal of the Royal Society of Arts*, lxxxvi, 1938, p. 926.

JACKS, G. V., and WHYTE, R. O. *Erosion and Soil Conservation*, Bulletin 25, Imperial Bureau of Pastures and Forage Crops, Aberystwyth, 1938.

—— —— *The Rape of the Earth: A World Survey of Soil Erosion*, London, 1939.

Soils and Men, Year Book of Agriculture, 1938, U.S. Dept. of Agr., Washington, D.C., 1938.

ALKALI SOILS

HILGARD, E. W. *Soils*, New York, 1906.

HOWARD, A. *Crop Production in India*, Oxford University Press, 1924.

KING, F. H. *Irrigation and Drainage*, London, 1900.

OSSENDOWSKI, F. *Man and Mystery in Asia*, London, 1924.

RUSSELL, SIR JOHN. *Soil Conditions and Plant Growth*, London, 1937.

CHAPTER XI

THE RETREAT OF THE CROP AND THE ANIMAL BEFORE THE PARASITE

IN the previous chapter we have seen how Nature, by means of soil erosion, removes any area of worn-out land and recreates new soil in a fresh place. Mismanagement of the land is followed later on by a New Deal, as it were, somewhere else. A similar rule applies to crops: the diseased crop is quietly but effectively labelled prior to removal for the manufacture of humus, so that the next generation of plants may benefit.

Mother earth has provided a vast organization for indicating the inefficient crop. Where the soil is infertile, where an unsuitable variety is being grown, or where some mistake has been made in management, Nature at once registers her disapproval through her Censors' Department. One or more of the groups of parasitic insects and fungi—the organisms which thrive on unhealthy living matter—are told off to point out that farming has failed. In the conventional language of to-day the crop is attacked by disease. In the writings of the specialist, a case has arisen for the control of a pest: a crop must be protected.

In recent years another form of disease—known as virus disease—has made its appearance. There is no obvious parasite in virus diseases, but insects among other agencies are able to transmit the trouble from diseased to apparently healthy plants in the neighbourhood. When the cell contents of affected plants are examined, the proteins exhibit definite abnormalities, thereby suggesting that the work of the green leaf is not effective; the synthesis of albuminoids seems to be incomplete. With the development of special research laboratories, like that at Cambridge, more and more of these diseases are being discovered and a considerable literature on the subject has arisen.

The virus diseases do not complete the story. A certain number of maladies occur in which the apparent cause is neither a fungus, an insect, nor a virus. These are grouped under the general title—physiological diseases: troubles arising from the collapse of the normal metabolic processes.

How has agricultural science dealt with the diseases of crops? The answer is both interesting and illuminating. The subject has been approached in a variety of ways, which can be briefly summed up under the following four heads:

1. The study of the life history of the pest, including the general relation of the parasite to the crop and the influence of the environment on the struggle for supremacy between the two. The main object of these investigations has been to discover some possible weakness in the life history of the pest which can be utilized to destroy it or to protect the plant from infection. An impressive volume of specialist literature has resulted. As the number of investigators grows and as their inquiries become more exhaustive and tend to cover a rapidly increasing proportion of the earth's surface, there is a corresponding increase in the volume of print. It is now almost impossible to take up any of the periodicals dealing with agricultural research without finding at least one long illustrated article describing some new disease. So vast has the literature become that the specialists themselves are unable to cope with it. Most of it can only be read by the workers in abstract, for which again new agencies have been created in the British Empire—the Imperial Bureaux of Entomology and Mycology—bodies which act as clearing-houses of information and deal with the published papers in a way reminiscent of the methods of the Banker's Clearing House in dealing with cheques.

2. The study of the natural parasites of insect pests, the breeding of these animals, and their actual introduction whenever this procedure promises success. A separate institution for this purpose has been founded at Farnham Royal in Buckinghamshire.

3. The protection of the crop from the inroads of the parasite. As a rule this takes two forms: (1) the discovery of insecticides and fungicides and the design of the necessary machinery for covering the crop with a thin film of poison which will destroy the parasite in the resting stage or before it can gain entry to the host; (2) the destruction of the parasite by burning, by the use of corrosive liquids like strong sulphuric acid, or by germicides added to the soil so that the amount of infecting material will be negligible.

4. The framing and conduct of regulations to protect an area from some foreign pest which has not yet made its appearance.

These follow the usual methods of quarantine. Importation of plants and seeds is prohibited altogether, introduction is permitted under licence, or the plant material is inspected and fumigated at the port of entry. The principle in all cases is the same—the crops must be protected from chance infection by some foreign parasite which might cause untold damage. As traffic by land, sea, and air grows in volume and becomes speeded up, it will be increasingly difficult to enforce these regulations. It is impossible even now to inspect all luggage and all merchandise and to prevent the smuggling of small packets of seed or cuttings of living plants. Indeed, if an investigation were to be made of the personal effects of the coolies passing backwards and forwards between India and Burma, India and the Federated Malay States and Ceylon, it would be seen what an extraordinary collection of articles these men and women carry about and how frequently plants and seeds are included. Enthusiasts in gardening often collect plants on their travels which interest them. The population, live stock, and factories of Great Britain are partly supplied with seeds from all over the world. By one or other of these agencies a few new pests are almost certain from time to time to enter the country. These quarantine methods therefore can never succeed.

More than fifty years have passed since the modern work on the diseases of plants first began. What has been the general result of all this study of vegetable pathology? Has it provided anything of permanent value to agriculture? Is the game worth the candle? Must agricultural science go on discovering more and more new pests and devising more and more poison sprays to destroy them or is there any alternative method of dealing with the situation? Why is there so much of this disease? Can the growing tale of the pests of Western agriculture be accounted for by some subtle change in practice? Can the cultivators of the East, for example, teach us anything about diseases and their control?

In this chapter an attempt will be made to answer these interesting questions.

It is a well-understood principle in business that any organization like agricultural research, which has grown by accretion rather than by the development of a considered plan, stands in need of a periodical critical examination to ascertain whether the results obtained correspond with the cost and whether any modifications are needed in the light of new knowledge and experience. I began such an investigation of the plant and animal disease section of agricultural science in 1899 and have steadily pursued it since. After forty years' work I feel sufficiently confident of my general conclusions to place them on record, and to ask for them to be considered on their merits.

I took up research in agriculture as a mycologist in the West Indies in 1899, where I specialized in the diseases of sugar-cane and cacao and became interested in tropical agriculture. Almost at once I discerned a fundamental weakness in the research organization: the mycologist had no land on which he could take his own advice about remedies before asking planters to adopt them.

My next post was botanist at Wye College in Kent, where I was in charge of the experiments on hops and had ample opportunities for studying the insect and fungous diseases of this interesting crop. But again I had no land on which I could try out certain ideas that were fermenting in my mind about the prevention of hop diseases. I observed one interesting thing: the increase in the resisting power to infection of the young hop flower which resulted from pollination. This observation has since brought about a change in the local practice: the male hop is now cultivated and ample pollination of the female flowers—the hops of commerce—occurs.

In 1905 I was appointed Imperial Economic Botanist to the Government of India. At the Pusa Agricultural Research Institute, largely through the support of the Director, the late Mr. Bernard Coventry, I had for the first time all the essentials for work—interesting problems, money, freedom, and last but not least, 75 acres of land on which I could grow crop in my own way and study their reaction to insect and fungous pests and other things. My real training in agricultural research then began—six years after leaving the University and obtaining all the paper qualifications and academic experience then needed by an investigator.

At the beginning of this second and intensive phase of my training, I resolved to break new ground and try out an idea (which first occurred to me in the West Indies), namely, to observe what happened when insect and fungous diseases were left alone and allowed to develop unchecked, and where indirect methods only, such as improved cultivation and more efficient varieties, were employed to prevent attack. This point of view derived considerable impetus from a preliminary study of Indian agriculture. The crops grown by the cultivators in the neighbourhood of Pusa were remarkably free from pests of all kinds; such things as insecticides and fungicides found no place in this ancient system of agriculture. I decided that I could not do better than watch the operations of these peasants, and acquire their traditional knowledge as rapidly as possible. For the time being, therefore, I regarded them as my professors of agriculture. Another group of instructors were obviously the insects and fungi themselves. The methods of the cultivators if followed would result in crops practically free from disease; the insects and fungi would be useful for pointing out unsuitable varieties and methods of farming inappropriate to the locality.

It was possible for me to approach the subject of plant diseases in this unorthodox manner for two reasons. In the first place the Agricultural Research Institute at Pusa was little more than a name when I arrived in India in 1905. Everything was fluid; there was nothing in the nature of an organized system of research in existence. In the second place, my duties, fortunately for me, had not been clearly defined. I was therefore able to break new ground, to widen the scope of economic botany until it became crop production, to base my investigations on a first-hand knowledge of Indian agriculture, and to take my own advice before offering it to other people. In this way I escaped the fate of the majority of agricultural investigators—the life of a laboratory hermit devoted to the service of an obsolete research organization. Instead, I spent my first five years in India ascertaining by practical experience the principles underlying health in crops.

In order to give my crops every chance of being attacked by parasites, nothing was done in the way of prevention; no insecticides and fungicides were used; no diseased material was ever destroyed. As my understanding of Indian agriculture progressed, and as my practice improved, a marked diminution of disease occurred. At the end of five years' tuition under my new professors—the peasants and the pests—the attacks of insects and fungi on all crops, whose root systems were suitable to the local soil conditions, became negligible. By 1910 I had learnt how to grow healthy crops, practically free from disease, without the slightest help from mycologists, entomologists, bacteriologists, agricultural chemists, statisticians, clearing-houses of information, artificial manures, spraying machines, insecticides, fungicides, germicides, and all the other expensive paraphernalia of the modern Experiment Station.

I then posed to myself the principles which appeared to underlie the diseases of plants:

1. Insects and fungi are not the real cause of plant diseases but only attack unsuitable varieties or crops imperfectly grown. Their true role is that of censors for pointing out the crops that are improperly nourished and so keeping our agriculture up to the mark. In other words, the pests must be looked upon as Nature's professors of agriculture: as an integral portion of any rational system of farming.

2. The policy of protecting crops from pests by means of sprays, powders, and so forth is unscientific and unsound as, even when successful, such procedure merely preserves the unfit and obscures the real problem—how to grow healthy crops.

3. The burning of diseased plants seems to be the unnecessary destruction of organic matter as no such provision as this exists in Nature, in which insects and fungi after all live and work.

This preliminary exploration of the ground suggested that the birthright of every crop is health, and that the correct method of dealing with disease at an Experiment Station is not to destroy the parasite, but to make use of it for tuning up agricultural practice.

Steps were then taken to apply these principles to oxen, the power unit in Indian agriculture. For this purpose it was necessary to have the work cattle under my own charge, to design their accommodation, and to arrange for their feeding, hygiene, and management. At first this was refused, but after persistent importunity, backed by the powerful support of the Member of the Viceroy's Council in charge of agriculture (the late Sir Robert Carlyle, K.C.S.I.), I was allowed to have charge of six pairs of oxen. I had little to learn in this matter as I belong to an old

M

agricultural family and was brought up on a farm which had made for itself a local reputation in the management of cattle. My work animals were most carefully selected and everything was done to provide them with suitable housing and with fresh green fodder, silage, and grain, all produced from fertile land. I was naturally intensely interested in watching the reaction of these well-chosen and well-fed oxen to diseases like rinderpest, septicaemia, and foot-and-mouth disease which frequently devastated the country-side.[1] None of my animals were segregated; none were inoculated; they frequently came in contact with diseased stock. As my small farm-yard at Pusa was only separated by a low hedge from one of the large cattle-sheds on the Pusa estate, in which outbreaks of foot-and-mouth disease often occurred, I have several times seen my oxen rubbing noses with foot-and-mouth cases. Nothing happened. The healthy well-fed animals reacted to this disease exactly as suitable varieties of crops, when properly grown, did to insect and fungous pests—no infection took place.

As the factors of time and place are important when testing any agricultural innovation, it now became necessary to try out the three principles referred to above over a reasonably long period and in new localities. This was done during the next twenty-one years at three centres: Pusa (1910–24), Quetta (summers of 1910 to 1918), and Indore (1924–31).

At Pusa, during the years 1910 to 1924, outbreaks of plant diseases were rare, except on certain cultures with deep root systems which were grown chiefly to provide a supply of infecting material for testing the disease resistance of new types obtained by plant-breeding methods. Poor soil aeration always encouraged disease at Pusa. The unit species of *Lathyrus sativus* provided perhaps the most interesting example of the connexion between soil aeration and insect attack. These unit species fell into three groups: surface-rooted types always immune to green-fly; deep-rooted types always heavily infected; types with intermediate root system always moderately infected. These sets of cultures were grown side by side year after year in small oblong plots about 10 feet wide. The green-fly infection repeated itself each year

---

[1] These epidemics are the result of starvation, due to the intense pressure of the bovine population on the limited food-supply.

and was determined not by the presence of the parasite, but by the root development of the host. Obviously the host had to be in a certain condition before infection could take place. The insect, therefore, was not the cause but the consequence of something else.

One of the crops under study at Pusa was tobacco. At first a great many malformed plants—since proved to be due to virus—made their appearance in my cultures. When care was devoted to the details of growing tobacco seed, to the raising of the seedlings in the nurseries, to transplanting and general soil management, this virus disease disappeared altogether. It was very common during the first three years; it then became infrequent; between 1910 and 1924 I never saw a single case. Nothing was done in the way of prevention beyond good farming methods and the building up of a fertile soil. I dismissed it at the time as one of the many mare's nests of agricultural science—things which have no real existence.

For a period of eight years, I was provided with a subsidiary experiment station on the loess soils of the Quetta valley for the study of the problems underlying fruit-growing and irrigation. I observed no fungous disease of any importance in the dry climate of the Quetta valley during the eight summers I spent there. In the grape gardens, run by the tribesmen on the well-drained slopes of the valley, I never observed any diseases—insect or fungous—on the grapes or on the vines, although they were planted on the floors of deep trenches, allowed to climb up the earth walls and were frequently irrigated. At first sight, all the conditions for mildew seemed to have been provided, yet I never saw a single speck. Three favourable factors probably accounted for this result. The climate was exceedingly dry, with considerable air movement and cloudless skies; the soil made use of by the roots of the vine was open, well drained, and exceptionally well aerated; the only manure used was farm-yard manure. Growth, yield, quality, and disease resistance left nothing to be desired.

The chief pest of fruit-trees at Quetta was green-fly soon after the young leaves appeared. This could be produced or avoided at will by careful attention to cultivation and irrigation. Any interference with soil aeration brought on this trouble; anything which promoted soil aeration prevented it. I frequently produced a strong attack of green-fly on peaches and almonds by over-irrigation during the winter and spring, and then stopped it dead by deep cultivation. The young shoots were covered with the pest below, but the upper portions of the same shoots were completely healthy. The green-fly never spread from the lower to the upper leaves on the same twig. The tribesmen got over the tendency of these loess soils to pack under irrigation in a very simple way. Lucerne was always grown in the fruit orchards, and regularly top-dressed with farm-yard manure. In this way the porosity of the soil was maintained and the green-fly kept in check.

At the Institute of Plant Industry, Indore, only two cases of disease occurred during the eight years I was there. The first occurred on a small field of gram (*Cicer arietinum*), about two-thirds of which was flooded for a few days one July, due to the temporary stoppage of one of the drainage canals which took storm water from adjacent areas through the estate. A map of the flooded area was made at the time. In October, about a month after sowing. this plot was heavily attacked by the gram caterpillar, the insect-infected area corresponding exactly with the inundation area. The rest of the plot escaped infection and grew normally. The insect did not spread to the other 50 acres of gram grown that year alongside. Some change in the food of the caterpillar had obviously been brought about by the alteration in the soil conditions caused by the temporary flooding. The second case of disease occurred in a field of *san* hemp (*Crotalaria juncea* L.) intended for green-manuring; however, this was not ploughed in but was kept for seed. After flowering the crop was smothered by a mildew; no seed was harvested. To produce a crop of seed of *san* on the black soils, it is necessary to manure the land with humus or farm-yard manure, when no infection takes place and an excellent crop of seed is obtained.

One experiment with cotton unfortunately could not be arranged in spite of all my efforts. At Indore there was a remarkable absence of all insect and fungous diseases of cotton. Good soil management, combined with dressings of humus, produced crops which were practically immune to all the pests of cotton. When the question of protecting India from the various cotton boll-worms and boll-weevils from America was discussed, I offered to have these let loose among my cotton cultures at Indore in order (1) to settle the question whether these troubles in the U.S.A. were due to the insect or to the way the cotton was grown, and (2) to subject my farming methods to a crucial test. I am pretty certain the insects would have found my cotton cultures very indifferent nourishment. My proposal, however, did not find favour with the entomological advisers of the Indian Cotton Committee and the matter dropped.

At Quetta and Indore there was no case of infectious disease among the oxen. The freedom from disease observed at Pusa was again experienced in the new localities—the Western Frontier and Central India.

It was soon discovered in the course of this work that the thing that matters most in crop production is a regular supply of well-made farm-yard manure and that the maintenance of soil fertility is the basis of health.

HUMUS AND DISEASE RESISTANCE

Even on the Experiment Stations the supply of farm-yard manure was always insufficient. The problem was how to increase it in a country where a good deal of the cattle-dung has to be burnt for fuel. The solution of this problem was suggested by the age-long practices of China. It involved the study of how best to convert the animal and vegetable wastes into humus, so that every holding in India could become self-supporting as regards manure. Such a problem did not fall within my sphere of work—the improvement of crops. It obviously necessitated a great deal of chemical work under my personal control. The organization of research at Pusa had gradually become more rigid; the old latitude which existed in the early days became a memory. That essential freedom, without which no progress is possible, had been gradually destroyed by the growth of a research organization based on fragments of science rather than on the practical problems which needed investigation. The instrument became more important than its purpose. Such organizations can only achieve their own destruction. This was the reason why I decided to leave Pusa and found a new centre where I should be free to follow the gleam unhampered and undisturbed. After a delay of six years, 1918 to 1924, the Indore Institute was founded. In due course a simple method, known as the Indore Process, of composting vegetable and animal wastes was devised, tested, and tried out on the 300 acres of land at the disposal of the Institute of Plant Industry, Indore. In a few years production more than doubled: the crops were to all intents and purposes immune from disease.

Since 1931 steps have been taken to get the Indore Process taken up in a number of countries, especially by the plantation industries such as coffee, tea, sugar, sisal, maize, cotton, tobacco, and rubber. The results obtained have already been discussed. In all these trials the conversion of vegetable and animal wastes into humus has been followed by a definite improvement in the health of the crops and of the live stock. My personal experience in India has been repeated all over the world. At the same time a number of interesting problems have been unearthed. One example will suffice. In Rhodesia humus protects the maize crop from the attacks of the witchweed (*Striga lutea*). Is this infestation a consequence of malnutrition? Is immunity conferred by the establishment of the mycorrhizal association? Answers to these questions would advance our knowledge and would suggest a number of fascinating problems for investigation.

THE MYCORRHIZAL ASSOCIATION AND DISEASE

Why is humus such an important factor in the health of the crop? The mycorrhizal association provides the clue. The steps by which this conclusion was reached in the case of tea have already been stated (p. 56).

This association is not confined to one particular forest crop. It occurs in most if not all our cultivated plants. During 1938 Dr. Rayner and Dr. Levisohn examined and reported on a large number of my samples—rubber, coffee, cacao, leguminous shade trees, green-manure crops, coco-nuts, tung, cardamons, vine, banana, cotton, sugar-cane, hops, strawberries, bulbs, grasses and clovers and so forth. In all of these the mycorrhizal association occurs. It is probably universal. We appear to be dealing with a very remarkable example of symbiosis in which certain soil fungi directly connect the humus in the soil with the roots of the crop.

This fungous tissue may contain as much as 10 per cent. of nitrogen in the form of protein, which is digested in the active roots and probably carried by the transpiration current to the seat of carbon assimilation in the green leaves. Its effective presence in the roots of the plant is associated with health; its absence is associated with diminished resistance to disease. Clearly the first step in investigating any plant disease in the future will be to see that the soil is fertile and that this fungous association is in full working order. If it is as important as is now suggested, there will be a marked improvement in the behaviour of the host once the fertility of the soil is restored. If it has no significance, a fertile soil will make no difference.

I have just obtained confirmatory results which prove how important humus is in helping a mycorrhiza-former—the apple—to throw off disease. In 1935 I began the restoration, by means of humus, of my own garden, the soil of which was completely worn-out when I acquired it in the summer of 1934. The apple trees were literally smothered with American blight, green-fly, and fruit-destroying caterpillars like the codlin moth. The quality of the fruit was poor. Nothing was done to control these pests beyond the gradual building up of the humus content of the soil. In three years the parasites disappeared; the trees were transformed; the foliage and the new wood now leave nothing to be desired; the quality of the fruit is first class. These trees will now be used for infection experiments in order to ascertain whether the fertility of the soil has been completely restored or not. The reaction of the trees to the various pests of the apple will answer this question. No soil analysis can tell me as much as the trees will.

The meaning of all this is clear. Nature has provided a marvellous piece of machinery for conferring disease-resistance on the crop. This machinery is only active in soil rich in humus; it is inactive or absent in infertile land and in similar soils manured with chemicals. The fuel needed to keep this machinery in motion is a regular supply of freshly prepared humus, properly made. Fertile soils then yield crops resistant to disease. Worn-out soils, even when stimulated with chemical manures, result in produce which needs the assistance of insecticides and fungicides to yield a crop at all. These in broad outline are the facts.

The complete scientific explanation of the working of this remarkable example of symbiosis remains to be provided. It would appear that in the mycorrhizal association Nature has given us a mechanism far more important and far more universal than the nodules on the roots of the clover family. It reconciles at one bound science and the age-long experience of the tillers of the soil as to the supreme importance of humus. There has always been a mental reservation on the part of the best farmers as to the value of artificial manures compared with good old-fashioned muck. The effect of the two on the soil and on the crop is never quite the same. Further, there is a growing conviction that the increase in plant and animal diseases is somehow connected with the use of artificials. In the old days of mixed farming the spraying machine was unknown, the toll taken by troubles like foot-and-mouth disease was insignificant compared with what it is now. The clue to all these differences—the mycorrhizal association— has been there all the time. It was not realized because the experiment stations have blindly followed the fashion set by Liebig and Rothamsted in thinking only of soil nutrients and have forgotten to look at the way the plant and the soil come into gear. An attempt has been made to apply science to a biological problem by means of one fragment of knowledge only.

THE INVESTIGATIONS OF TO-MORROW

The next step in this investigation is to test the soundness of the views put forward. This has been started by composting diseased material and then using the humus to grow another crop on the same land. Diseased tomatoes have been converted into humus by one of the large growers in the south of England and the compost has been used to grow a second crop in the same greenhouses. No infection occurred.

The final proof that insects, fungi, viruses and so forth are not the cause of disease will be provided when the infection experiments of to-morrow are undertaken. Instead of conducting these trials on plants and animals grown anyhow, the experimental material will be plants and animals, properly selected, efficiently managed, and nourished by or on the produce of a fertile soil. Such plants can be sprayed with active fungous and insect material without harm. Among such herds of cattle cases of foot-and-mouth disease can be introduced without any danger of serious infection. The afflicted animals themselves will recover. When some audacious innovator of the Hosier type, who has no interest in the maintenance of the existing research structure, conducts such experiments, the vast fabric of disease-control which has been erected in countries like Great Britain will finally collapse. Farmers will emancipate themselves from the thraldom created by the fear of the parasite. Another step forward will be taken . which will not stop at farming.

My self-imposed task is approaching completion. I have examined in great detail for forty years the principles underlying the treatment of plant and animal diseases, as well as the practice based on these principles. It now remains to sum up this experience and to offer suggestions for the future.

There can be no doubt that the work in progress on disease at the Experiment Stations is a gigantic and expensive failure, that its continuance on present lines can lead us nowhere and that steps must be taken without delay to place it on sounder lines.

The cause of this failure is not far to seek. The investigations have been undertaken by specialists. The problems of disease have not been studied as a whole, but have been divorced from practice, split up, departmentalized and confined to the experts most conversant with the particular fragment of science which deals with some organism associated with the disease.

This specialist approach is bound to fail. This is obvious when we consider: (1) the real problem—how to grow healthy crops and how to raise healthy animals, and (2) the nature of disease—the breakdown of a complex biological system, which includes the soil in its relation to the plant and the animal. The problem must include agriculture as an art. The investigator must therefore be a farmer as well as a scientist, and must keep simultaneously in mind all the factors involved. Above all he must be on his guard to avoid wasting his life in the study of a mare's nest: in dealing with a subject which owes its existence to bad farming which will disappear the moment sound methods of husbandry are employed.

The problems presented by the retreat of the crop and of the animal before the parasite and the conventional methods of investigation of these questions are clearly out of relation. It follows therefore that a research organization which has lost direction and has permitted such a state of things to arise and to develop must itself be in need of overhaul. This task has been attempted; the existing structure of agricultural research has been subjected to a critical examination; the results are set out in Chapter XIII.

BIBLIOGRAPHY

HOWARD, A. *Crop Production in India*, Oxford University Press, 1924, p. 176.
—— 'The Role of Insects and Fungi in Agriculture', *Empire Cotton Growing Review*, xiii, 1936, p. 186.
—— 'Insects and Fungi in Agriculture', *Empire Cotton Growing Review*, xv, 1938, p. 215.
TIMSON, S. D. 'Humus and Witchweed', *Rhodesia Agricultural Journal*, xxxv, 1938, p. 805.

CHAPTER XII

SOIL FERTILITY AND NATIONAL HEALTH

IN the last chapter the retreat of the crop and the animal before the parasite was discussed. Disease was regarded as Nature's verdict on systems of agriculture in which the soil is deprived of its manurial rights. When the store of humus is used up and not replenished, both crops and animals first cease to thrive and then often fall a prey to disease. In other words, one of the chief causes of disease on the farm is bad soil management.

How does the produce of an impoverished soil affect the men and women who have to consume it? This is the theme of the present chapter. It is discussed, not on the basis of complete results, but from the point of view of a very promising hypothesis for future work. No other presentation is possible because of the paucity, for the moment, of direct evidence and the natural difficulties of the subject.

In the case of crops and live stock, experiments are easy. The investigator is not hampered in any way; he has full control of his material and freedom in experimentation. He cannot experiment on human beings in the same way. The only subjects that might conceivably be used for nutrition experiments on conventional lines are to be found in concentration camps, in convict prisons, and in asylums. Objections to using them for such purposes would almost certainly be raised. Even if they were not, the investigator would be dealing with life in captivity and with abnormal conditions. Any results obtained would not necessarily apply to the population as a whole.

Perhaps the chief difficulty at the moment in following up the possible connexion between the produce of a fertile soil and the health of the people who have to consume it, is to obtain from well-farmed land regular supplies of such produce in a perfectly fresh condition. Except in a few cases, food is not marketed according to the way it is grown. The buyer knows nothing of how it was manured. The only way to obtain suitable material would be for the investigator to take up a piece of land and grow the food itself. This, so far as my knowledge goes, has not been done. This omission alone explains the scarcity of direct results and why so little real progress has been made in human nutrition. Most of the work of the past has been founded on the use of food material very indifferently grown. Moreover, no particular care has been taken to see that the food has been eaten fresh from its source. Such investigations therefore can have no solid foundation.

Apart from the evidence that can be gathered from nutrition experiments, is there anything to be learnt about health from agriculture itself? Can the East which, long before the Roman Empire began or America was discovered, had already developed the systems of good farming which are in full swing to-day, throw any light on the relation between a fertile soil and a healthy population? It is well known that both China and India can show large areas of well-farmed land, which for centuries have carried very large populations. Unfortunately two factors—over-population and periodic crop failures due to irregular rainfall—make it almost impossible to draw any general lessons from these countries. The population, looked at in the mass, is always re-covering from one catastrophe after another. Over-population introduces a disturbing factor—long-continued semi-starvation—so powerful in its effects on the race and on the individual that any benefits arising from a fertile soil are entirely obscured.

When, however, the population of the various parts of India is examined, very suggestive differences between the races which make up its 350 million inhabitants are disclosed. The physique seen in the northern area is strikingly superior to that of the southern, eastern, and western tracts. We owe the investigation of the causes of these differences to McCarrison, who found that they corresponded with the food consumed. There is a gradually diminishing value in the food from the north to the east, south and west in respect of the amount and quality of the proteins, the quality of the cereal grains forming the staple article of diet, the quantity and quality of the fats, the mineral and vitamin content, as well as in the balance of the food as a whole.

Generally speaking the people of northern India, which include some of the finest races of mankind, are wheat eaters, the wheat being consumed in the form of thin, flat cakes made from flour, coarse but fresh-ground in a quern. All the proteins, vitamins, and mineral salts in the grain are consumed. The second most important article of diet is fresh milk and milk products—clarified butter, curds, and buttermilk; the third item is the seed of pulse crops; the fourth vegetables and fruit. Meat, as a rule, is very sparingly eaten except by the Pathans.

Turn now to the other parts of India, east, west, and south, in which the rice tracts provide the staple food. This cereal—a relatively poor grain at best—is parboiled, milled or polished, washed in many changes of water, and finally boiled. It is thereby deprived of much of its protein and mineral salts and of almost all its vitamins. In addition, very little milk or milk products are consumed, while the protein content of the diet is low both in amount and quality. Vegetables and fruit are only sparingly eaten. It is these shortcomings in their food that explain the poor physique of the peoples of the rice areas.

In order to prove that these bodily differences were due to food, McCarrison carried out experiments on young growing rats. When young growing rats of healthy stock were fed on diets similar to those of the races of northern India, the health and physique of the rats were good; when they were fed on the diets in vogue in the rice areas, the health and physique of the rats were bad; when they were fed on the diets of races with middling physique, the health and physique of the rats were middling. Other things being equal, good or bad diet led to good or bad health and physique.

When the health and physique of the various northern Indian races were studied in detail the best were those of the Hunzas, a hardy, agile, and vigorous people, living in one of the high mountain valleys of the Gilgit Agency, where an ancient system of irrigated terraces has been maintained for thousands of years in a high state of fertility. There is little or no difference between the kinds of food eaten by these hillmen and by the rest of northern India. There is, however, a great difference in the way these foods are grown. The total area of the irrigated terraces of the Hunzas is small; ample soil aeration results from their construction; the irrigation water brings annual additions of fine silt produced by the neighbouring glacier; the very greatest care is taken to return to the soil all human, animal, and vegetable wastes after being first composted together. Land is limited: upon the way it is looked after life depends. A perfect agriculture, in which all the factors that combine to produce high quality in food, naturally results.

What of the people who live upon this produce? In *The Wheel of Health*, Wrench has gathered together all the information available and has laid stress on their marvellous agility and endurance, good temper and cheerfulness. These men think nothing of covering the 60 miles to Gilgit on foot in one stretch, doing their business and then returning.

There is one point about the Hunza agriculture which needs further investigation. The staircase cultivation of these hillmen receives annual dressings of fresh rock-powder, produced by the grinding effect of the glacier ice on the rocks and carried to the fields in the irrigation water. Is there any benefit conferred on the soil and on the plant by these annual additions of finely divided materials? We do not know the composition of this silt. If it contains finely divided limestone its value is obvious. If it is made up for the most part of crushed silicates, its possible significance awaits investigation. Do the mineral residues in the soil need renewal as humus does? If so, then Nature has provided us with an Experiment Station ready-made and with results that cannot be neglected. Perhaps in the years to come, some heaven-sent investigator of the Charles Darwin type will go thoroughly into this Hunza question on the spot, and will set out clearly all the factors on which their agriculture and their marvellous health depend.

A study of the races of India and of their diet, coupled with the experimental work on rats carried out by McCarrison, leaves no doubt that the greatest single factor in the production of good health is the right kind of food and the greatest single factor in the production of bad health is the wrong kind of food. Further, the very remarkable health and physique enjoyed by the Hunza hillmen appears to be due to the efficiency of their ancient system of farming.

These results suggest that the population of Great Britain should be studied and the efficiency of our food supply investigated. If the physique and health of a nation ultimately depend on the fresh

produce of well-farmed land, if bad farming is a factor in the production of poor physique and bad health, we must set about
improving our agriculture without delay.

Two very different methods have recently been employed for
testing the efficiency of the food supply of this country. In the
first case (Cheshire) the population of a whole county, which
includes both rural and urban areas, has been studied in the mass
for a period of twenty-five years and the general results have been
recorded. In the second case (Peckham) a number of families
have been periodically examined with a view to throwing light
on the general health and efficiency of a group of comparatively
well-to-do workers in a city like London.

The methods adopted in the study of the population of Cheshire
and in the publication of the results are highly original. About
twenty-five years ago, the National Health Insurance Act for the
Prevention and Cure of Sickness came into force. This measure
has brought the population under close medical observation for a
quarter of a century. If the experience of the Panel and family
doctors of the county could be synthesized, valuable information
as to the general health of the community would be available.
This has been accomplished. The local Medical and Panel Committee of Cheshire, which is in touch with the 600 family doctors
of the county, has recorded its experience in the form of a Medical
Testament. They find it possible to report definite progress in the
'Cure of Sickness'—the second part of the objective framed by the
National Health Insurance Act. There is no doubt that we have
learnt to 'postpone the event of death' and this is the more
remarkable in view of the rise in sickness, in short the failure of the
first part of the Act's objective. On this latter count there is no
room for complacency.

'Our daily work brings us repeatedly to the same point—this
illness results from a life-time of wrong nutrition.' They then
examine the consequences of wrong nutrition under four heads—
bad teeth, rickets, anaemia, and constipation—and indicate how
all this and much other trouble can be prevented by right feeding.
For example, in dealing with the bad teeth of English children
some striking facts are emphasized. In 1936 out of 3,463,948
schoolchildren examined, no less than 2,425,299 needed dental

treatment. That this reproach can be removed has been shown by Tristan da Cunha, where the population is fed on the fresh produce of sea and soil—fish, potatoes, and seabirds' eggs are the staple diet with sufficient milk and butter, meat occasionally and some vegetables—all raised naturally without the help of artificial manures and poison sprays. In 1932, 156 persons examined had 3,181 permanent teeth of which 74 were carious. Imported flour and sugar have been brought in to a greater extent of late, which may account for the tendency of the teeth to deteriorate observed in 1937.

The Testament then goes on to the work of McCarrison (referred to above) 'whose experiments afford convincing proof of the effects of food and guidance in the application of the knowledge acquired'. This has been applied in local practice in Cheshire and the results have been amazing. Two examples will suffice.

1. In a Cheshire village the nutrition of expectant mothers is supervised by the local doctor once a month. The food of the mother is whole-meal bread, raw milk, butter, Cheshire cheese, oatmeal porridge, eggs, broth, salads in abundance, green leaf vegetables, liver and fish weekly, fruit in abundance and a little meat. The whole-meal bread is made from a mixture of two parts locally grown wheat, pulverized by a steel fan revolving 2,500 times a minute, and one part of raw wheat-germ (fresh off the rollers of a Liverpool mill). The flour is baked within thirty-six hours at most—a point to be rigidly insisted upon—and a rather close but very palatable bread is obtained. With rare exceptions the mothers feed their infants at the breast—nine months is advised and then very slow weaning finishing at about a year. The nursing mother's food continues as in pregnancy; the infants are fed five times a day with four-hour intervals beginning at 6 a.m. The children are splendid; perfect sets of teeth are now more common; they sleep well; pulmonary diseases are almost unknown; one of their most striking features is their good humour and happiness. They are sturdy-limbed, beautifully skinned, normal children. This was not a scientific experiment. It was part of the work of family practice. The human material was entirely unselected and the food was not specially grown; but that, in spite of these imperfections, the practical application of McCarrison's

work should yield recognizable results shows that in a single generation improvement of the race can be achieved.

2. A young Irishman aged 23, with a physique and an alertness of mind and body it was a delight to behold, was found to be suffering from catarrhal jaundice after two months' residence in England, where he had been living on a diet mostly composed of bacon, white bread, meat sandwiches and tea, with a little meat and an occasional egg. In Ireland his food had consisted of the fresh natural products of the soil—potatoes, porridge, milk and milk products, broth (made from vegetables) and occasional meat, eggs, and fish. The change over to a diet of white bread and sophisticated food was at once followed by disease. This case shows how quickly good health can be lost by improper feeding.

The Testament then leaves the purely medical domain and deals with that principle or quality in the varied diets which produces the same result—health and freedom from disease (such as the Esquimaux on flesh, liver, blubber, and fish; the Hunzas and Sikhs on wheaten chappattis, fruit, milk, sprouted legumes and a little meat; the islanders of Tristan on potatoes, seabirds' eggs, fish, and cabbage). In all these cases the diets have one thing in common—the food is fresh and little altered by preparation. The harvest of the sea is a natural product. When the foods are based on agriculture, the natural cycle from soil to plant, animal and man is complete without the intervention of any chemical or substitution phase. In other words, when the natural produce of sea and soil has escaped the attention of agricultural science and the various food preservation processes, it would seem that health results and that there is a marked absence of disease.

The last part of the Medical Testament deals with my own work on the connexion between a fertile soil and healthy plants and animals; with the means by which soil fertility can be restored and maintained; with a number of examples in which this has been done. These have already been described and need not be repeated.

This remarkable document concludes with the following words:

'The better manuring of the home land so as to bring an ample succession of fresh food crops to the tables of our people, the arrest of the present exhaustion of the soil and the restoration and permanent

N

maintenance of its fertility concern us very closely. For nutrition and the quality of food are the paramount factors in fitness. No health campaign can succeed unless the materials of which the bodies are built are sound. At present they are not.

'Probably half our work is wasted, since our patients are so fed from the cradle, indeed before the cradle, that they are certain contributions to a C3 nation. Even our country people share the white bread, tinned salmon, dried milk régime. Against this the efforts of the doctor resemble those of Sisiphus.

'This is our medical testament, given to all whom it may concern—and whom does it not concern?'

The Testament was put to a public meeting held at Crewe on March 22nd, 1939, by the Lord Lieutenant of Cheshire, Sir William Bromley-Davenport, and carried by a unanimous vote. More than five hundred persons representing the activities of the County of Cheshire were present. The Medical Testament was published in full in the issue of the *British Medical Journal* of April 15th, 1939. It has been widely noticed in the press all over the Empire.

The experience of the Cheshire doctors is supported by the work of Doctors Williamson and Pearse at the Peckham Health Centre in South London. In connexion with the study of families whose average wage is between £3. 15s. and £4. 10s. a week, about 20,000 medical examinations have been recorded. The results have recently been published in book form under the title *Biologists in Search of Material* (Faber & Faber). It was found that no less than 83 per cent. of apparently normal people had something the matter with them, ranging from some minor maladjustment to incipient disease. One of the most important contributions of these Peckham pioneers has been to unearth the beginnings of a C3 population. The next step will be to see how far these early symptoms of trouble can be removed by fresh food grown on fertile soil. For this the Centre must have: (1) a large area of land of its own on which vegetables, milk, and meat can be raised, and (2) a mill and bakehouse in which whole-meal bread, produced on Cheshire lines from English wheat grown on fertile soil, can be prepared. In this way a large amount of food resembling that of the Hunza hillmen can be obtained. The medical records of the families which consume this produce, after the change over from the canned stuff of the shops and the semi-carrion of the cold stores has been made, will form interesting reading.

The health of our population has been studied by these two very different methods. Both lead to the same conclusion, namely, that all is not well: that there is an enormous amount of indisposition, inefficiency, and actual disease. The Medical Testament boldly suggests that want of freshness in the food and improper methods of agriculture are at the root of the mischief. This provides a stimulating hypothesis for future work. A case for action has been established. The basis of the public health system of the future has been foreshadowed.

A certain amount of supporting evidence is already available. Two recent examples can be quoted, the first dealing with live stock, the second with schoolboys.

At Marden Park in Surrey, Sir Bernard Greenwell has found that a change over to a ration of fresh home-grown food (raised on soil manured with humus) fed to poultry and pigs has been followed by three important results: (1) the infantile mortality has to all intents and purposes disappeared; (2) the general health and well-being of the live stock has markedly improved; (3) a reduction of about 10 per cent. in the ration has been obtained because such home-grown produce possesses an extra-satisfying power.

At a large preparatory school near London, at which both boarders and day-boys are educated, the change over from vegetables, grown with artificial manures, to produce grown on the same land with Indore compost has been accompanied by results of considerable interest to parents and to the medical profession. Formerly, in the days when artificials were used, cases of colds, measles, and scarlet fever used to run through the school. Now they tend to be confined to the single case imported from outside. Further, the taste and quality of the vegetables have definitely improved since they were raised with humus.

Much more work on these lines is needed. A search will have to be made throughout Great Britain and Northern Ireland for resident communities such as boarding-schools, training centres, the resident staff of hospitals and convalescent homes which satisfy the following four conditions: (1) the control of sufficient fertile and well-farmed land for growing the vegetables, fruit, milk and milk products, and meat required by the residents; (2) a mill and bakehouse for producing wholemeal bread with the new Cambridge wheats grown on fertile soil without the assistance of artificial manures; (3) the medical supervision of the community by a carefully chosen disciple of preventive medicine; (4) a man or woman in control who is keenly interested in putting the findings of the Medical Testament to the test and who is prepared to surmount any difficulties that may arise. In a very few years it is more than probable that islands of health will arise in an ocean of indisposition. No controls will be necessary—these will be provided by the country-side round about. Elaborate statistics will be superfluous as the improved health of these communities will speak for itself and will need no support from numbers, tables, curves, and the higher mathematics. Mother earth in the appearance of her children will provide all that is necessary. The materials for Medical Testament No. 2 will then be available. Cheshire no doubt will again take the lead and provide a second milestone on the long road which must be traversed before this earth can be made ready to receive her children.

In this work research can assist. Medical investigations should be deflected from the sterile desert of disease to the study of health —to mankind in relation to his environment. Agricultural research, after reorganization on the lines suggested in the next chapter, should start afresh from a new base-line—soil fertility— and so provide the raw material for the nutritional studies of the future—fresh produce from fertile soil. The agricultural colleges with their farms should devote some of their resources to feeding themselves, and so demonstrating what the products of well-farmed land can accomplish. They should strive to equal and then to surpass what a tribe of northern India has already achieved.

BIBLIOGRAPHY

SCOTT WILLIAMSON, G., and INNES PEARSE, H. *Biologists in Search of Material*, Faber & Faber, London, 1938.
HOWARD, SIR ALBERT. 'Medical Testament on Nutrition', *British Medical Journal*, May 27th, 1939, p. 1106.
MCCARRISON, SIR ROBERT. 'Nutrition and National Health' (Cantor Lectures), *Journal of the Royal Society of Arts*, lxxxiv, 1936, pp. 1047, 1067, and 1087.
'Medical Testament on Nutrition', Supplement to the *British Medical Journal*, April 15th, 1939, p. 157; Supplement to the *New English Weekly*, April 6th, 1939.
WRENCH, G. T. *The Wheel of Health*, London, 1938.

PART IV

AGRICULTURAL RESEARCH

CHAPTER XIII

A CRITICISM OF PRESENT-DAY AGRICULTURAL RESEARCH

A WANT of relation between the conventional methods of investigation and the nature of disease in plants and animals has been shown to exist. The vast fabric of agricultural research must now be examined in order to determine whether effective contact has been maintained with the problems of farming. This is the theme of the present chapter.

The application of science to agriculture is a comparatively modern development, which began in 1834 when Boussingault laid the foundations of agricultural chemistry. Previously all the improvements in farming practice resulted from the labours of a few exceptional men, whose innovations were afterwards copied by their neighbours. Progress took place by imitation. After 1834 the scientific investigator became a factor in discovery. The first notable advance by this new agency occurred in 1840, when Liebig's classical monograph on agricultural chemistry appeared. This at once attracted the attention of agriculturists. Liebig was a great personality, an investigator of genius endowed with imagination, initiative, and leadership and was exceptionally well qualified for the scientific side of his task—the application of chemistry to agriculture. He soon discovered two important things: (1) that the ashes of plants gave useful information as to the requirements of crops, and (2) that a watery extract of humus gave little or no residue on evaporation. As the carbon of the plant was obtained from the atmosphere by assimilation in the green leaf, everything seemed to point to the supreme importance of the soil and the soil solution in the raising of crops. It was only necessary to analyse the ashes of plants, then the soil, and to apply to the latter the necessary salts to obtain full crops. To establish the new point of view the humus theory, which then held the field, had to be demolished. According to this theory the plant fed on humus. Liebig believed he had shown that this view was untenable; humus was insoluble in water and therefore could not influence the soil solution.

In all this he followed the science of the moment. In his onslaught on the humus theory he was so sure of his ground that he did not call in Nature to verify his conclusions. It did not occur to him that while the humus theory, as then expressed, might be wrong, humus itself might be right. Like so many of his disciples in the years to come, he failed to attach importance to the fact that the surface soil always contains very active humus, and did not perceive that critical field experiments, designed to find out if chemical manures were sufficient to supply all the needs of crops, should always be done on the sub-soil, after removing the top 9 inches or so. If this is not arranged for, the yield of any crop may be influenced by the humus already in the soil. Failure to perceive this obvious fact is the main reason why Liebig and his disciples went astray.

He also failed to realize the supreme importance to the investigator of a first-hand knowledge of practical agriculture, and the significance of the past experience of the tillers of the soil. He was only qualified for his task on the scientific side; he was no farmer; as an investigator of the ancient art of agriculture he was only half a man. He was unable to visualize his problem from two very different points of view at one and the same moment—the scientific and the practical. His failure has cast its shadow on much of the scientific investigation of the next hundred years. Rothamsted, which started in 1843, was profoundly influenced by the Liebig tradition. The celebrated experiments on Broadbalk Field caught the fancy of the farming world. They were so telling, so systematic, so spectacular that they set the fashion till the end of the last century, when the great era of agricultural chemistry began to wane. During this period (1840–1900), agricultural science was a branch of chemistry; the use of artificial manures became firmly welded into the work and outlook of the Experiment Stations; the great importance of nitrogen (N), phosphorus (P), and potash (K) in the soil solution was established; what may briefly be described as the NPK mentality was born.

The trials of chemical manures, however, brought the investigators from the laboratory to the land; they came into frequent contact with practice; their outlook and experience gradually widened. One result was the discovery of the limitations of chemical science; the deficiencies of the soil, suggested by chemical analysis, were not always made up by the addition of the appropriate artificial manure; the problems of crop production could not be dealt with by chemistry alone. The physical texture of the soil began to be considered; the pioneering work of Hilgard and King in America led to the development of a new branch of the subject—soil physics—which is still being explored. Pasteur's work on fermentation and allied subjects, by drawing attention to the fact that the soil is inhabited by bacteria and other forms of life, disclosed a new world. A notable elucidation of the complex life of the soil was contributed by Charles Darwin's fascinating account of the earthworm. The organisms concerned with the nitrification of organic matter were discovered by Winogradsky and the conditions necessary for their activity in pure cultures were determined. Another branch of agricultural science—soil bacteriology—arose. While the biology and physics of the soil were being studied, a new school of soil science arose in Russia. Soils began to be regarded as independent natural growths: to have form and structure due to climate, vegetation, and geological origin. Systems of soil classification, based primarily on the soil profile, with an appropriate nomenclature developed in harmony with these views, which, for the moment, have been widely accepted. A new branch of soil science—pedology—arose. The Liebig conception of soil fertility was thus gradually enlarged and it became clear that the problem of increasing the produce of the soil did not lie within the domain of any one science but embraced at least four—chemistry, physics, bacteriology, and geology.

At the beginning of the present century, the investigators began to pay more attention to what is after all the chief agent in crop production—the plant itself. The rediscovery of Mendel's law by Correns, the conception of the unit species which followed the work of Johannsen and the recognition of its importance in improvement by selection have led directly to the modern studies of cultivated crops, in which the Russians have made such noteworthy contributions. The whole world is now being ransacked to provide the plant breeders with a wide range of raw material. These botanical investigations are constantly broadening and now embrace the root system, its relation to the soil type, the resistance of the plant to disease, as well as the internal mechanism by which inheritance takes place. The practical results of the last forty years which have followed the application of botanical science to agriculture are very considerable. In wheat, for example, the labours of Saunders in Canada led to the production of Marquis, an early variety with short straw, which soon covered 20,000,000 acres in Canada and the neighbouring States of the Union. This is the most successful wheat-hybrid yet produced. In Australia the new wheats raised by Farrer were soon widely cultivated. In England the new hybrids raised at Cambridge established themselves in the wheat-growing areas of this country. In India the Pusa wheats covered several million acres of land. By 1925 the total area of the new varieties of wheat had reached over 25,000,000 acres. When the annual dividend, in the form of increased wealth, was compared with the capital invested in these investigations, it was at once evident that the return was many times greater than that yielded by the most successful industrial enterprise. Similar results have been obtained with other crops. The new varieties of malting barley, raised by Beaven, have for years been a feature of the English country-side; the new varieties of sugar-cane produced by Barber at Coimbatore in South India soon replaced the indigenous types of cane in northern India. In cotton, jute, rice, grasses, and clovers and many other crops new varieties have been obtained; the old varieties are being systematically replaced. Nevertheless the gain per acre obtained by changing the variety is as a rule small. As will be seen in the next chapter, the great problem of agriculture at the moment is the intensive cultivation of these new types; how best to arrange a marriage between the new variety and a fertile soil. Unless this is done, the value of a new variety can only be transient; the increased yield will be obtained at the expense of the soil capital; the labours of the plant breeders will have provided another boomerang.

A number of other developments have taken place which must briefly be mentioned. Since the Great War the factories then engaged in the fixation of atmospheric nitrogen for the manufacture of the vast quantities of explosives, needed to defend and to destroy armies well entrenched, have had to find a new market. This was provided by the large area of land impoverished by the over-cropping of the war period. A demand was created by the low price at which the mass-produced unit of nitrogen could be put on the market and by the reliability of the product. Phosphates and potash fell into line. Ingenious mixtures of artificial manures, containing everything supposed to be needed by the various crops, could be purchased all over the world. Sales increased rapidly; the majority of farmers and market gardeners soon based their manurial programme on the cheapest forms of nitrogen, phosphorus, and potash or on the cheapest mixtures. During the last twenty years the progress of the artificial manure industry has been phenomenal; the age of the manure bag has arrived; the Liebig tradition returned in full force.

The testing of artificial manures and new varieties has necessitated innumerable field experiments, the published results of which are bewildering in their volume, their diversity, and often in the conclusions to be drawn from them. By a judicious selection of this material, it is possible to prove or disprove anything or everything. Something was obviously needed to regulate the torrent of field results and to ensure a greater measure of reliability. This was attempted by the help of mathematics. The technique was overhauled; the field plots were 'replicated' and 'randomized'; the figures were subjected to a rigid statistical scrutiny. Only those results which are fortunate enough to secure what has been described as the fastidious approval of the higher mathematics are now accepted. There is an obvious weakness in the technique of these field experiments which must be mentioned. Small plots and farms are very different things. It is impossible to manage a small plot as a self-contained unit in the same way as a good farm is conducted. The essential relation between live stock and the land is lost; there are no means of maintaining the fertility of the soil by suitable rotations as is the rule in good farming. The plot and the farm are obviously out of relation; the plot does not even represent the field in which it occurs. A collection of field plots cannot represent the agricultural problem they set out to investigate. It follows that any findings based on the behaviour of these small fragments of artificially manured land are unlikely to apply to agriculture. What possible advantage therefore can be obtained by the application of the higher mathematics to a technique which is so fundamentally unsound?

With the introduction of artificials there has been a continuous increase in disease, both in crops and in live stock. This subject has already been discussed (p. 156). It is mentioned again to remind the reader of the vast volume of research on this topic, completed and in progress.

Side by side with the intrusion of mathematics into agriculture, another branch of the subject has grown up—economics. The need for reducing expenditure so that farming could yield a profit has brought every operation, including manuring and the treatment of disease, under examination in order to ascertain the cost and what profit, if any, results. Costings are everywhere the rule; the value of any experiment and innovation is largely determined by the amount of profit which can be wrung from Mother earth. The output of the farm and of the factory have been looked at from the same standpoint—dividends. Agriculture joined the ranks of industry.

Agricultural science, like Topsy, has indeed grown. In little more than forty years a vast system of research institutes, experimental farms, and district organizations (for bringing the results of research to the farming community) has been created all over the world. As this research structure has grown up in piecemeal fashion as a result of the work of the pioneers, it will be interesting to examine it and to ascertain whether or not direction has been maintained. Has the present organization any virtue in itself or does it merely crystallize the stages reached in the scientific exploration of a vast biological complex? If it is useful it will be justified by results; if its value is merely historical, its reform can only be a question of time.

In Great Britain two documents[1] have recently appeared which make it easy to conduct an inquest on agricultural research in this country. They describe fully the structure and working of the

---

[1] *Constitution and Functions of the Agricultural Research Council*, H.M. Stationery Office, London, 1938; *Report on Agricultural Research in Great Britain*, PEP., 16 Queen Anne's Gate, London, S.W. 1, 1938.

official machine which controls and finances research, the organization of the work itself, and the methods of making the results known to farmers. In addition to the Treasury and the Committee of the Privy Council, official control is exercised by no less than three other organizations: (1) The Ministry of Agriculture (which administers the grants); (2) the Development Commission (which awards funds from grants placed at its disposal by the Treasury); and (3) the Agricultural Research Council (which reviews and advises on applications for grants, and also co-ordinates State-aided agricultural research in Great Britain). Eventually the Research Institutes, which carry out the work, are reached.

These Research Institutes are fifty in number and are of three types:

(a) Government laboratories or research stations;
(b) Institutes attached to universities or university colleges;
(c) Independent institutes.

Most of these institutes were set up in 1911 to provide for basic research in each of the agricultural sciences: agricultural economics, soil science, plant physiology, plant breeding, horticulture and fruit research, plant pathology, animal heredity and genetics, animal physiology and nutrition, animal diseases, dairy research, food preservation and transport, agricultural engineering and agricultural meteorology. These groups can again be divided into four classes: background research (dealing with fundamental scientific principles); basic research (the recognized sphere of the research institute); *ad hoc* research (the study of specific practical problems, as they arise, such as the control of foot-and-mouth disease); pilot or development research (such as the growing on of new strains of plants).

After research proper, the organization then deals with the results of its investigations. The first stage in this process is the Provincial Advisory Service which operates in sixteen provinces. From one to seven Advisory Officers are stationed at each centre, their specialized knowledge being at the disposal of County Organizers and farmers. The final link in the long chain from the Treasury to the soil is provided by the Agricultural Organizers of the County Councils, who act as a free Scientific Information Bureau for farmers and market gardeners. Most counties also support farm institutes which provide technical education and also have experimental farms of their own. Appended to this research structure are two Imperial Institutes and nine Imperial Bureaux, which provide an information and abstracting service in entomology, mycology, soil science, animal health, animal nutrition and genetics, plant genetics, fruit production, agricultural parasitology, and dairying. The number of agricultural research workers in Great Britain is about 1,000. The total State expenditure on agricultural research amounted in 1938 to about £700,000. This is about 90 per cent. of the total cost, the remaining 10 per cent. being met by local authorities, universities, marketing boards, private companies and individuals, agricultural societies, fees, and sales of produce. The farmers, even when organized as marketing boards, have shown little recognition of the value of research and make no serious contribution to its cost.

A formidable, complex, and costly organization has thus grown up since 1911. No less than seven organs of the Central Government have to do with agricultural research, the personnel of which has to be fed with a constant stream of reports, memoranda, and information which must absorb a large amount of the time and energy of the men who really matter—the investigators. A feature of the official control is the committee, a device which has developed almost beyond belief since the Agricultural Research Council came into being in 1934. Six standing committees were first formed to carry out a survey of existing research. These led to a crop of new committees to go farther into matters disclosed by this preliminary survey. In addition to the six standing committees, no less than fifteen scientific committees are dealing with the most important branches of research. Twelve of these fifteen committees are considering the diseases of crops and live stock —the main preoccupation of the Council at the present time.

Is so much machinery necessary? Between the Treasury (which decides what sum can be granted) and the Research Institutes, would not a single agency such as the Ministry of Agriculture be all that is needed in the way of control? This would appear likely when it is remembered that there is one thing only in research that matters—the man or woman who is to undertake it. Once these are found and provided with the means, nothing else is necessary. The best service the official organization can then perform is to remain in the background, ready to help when the workers need assistance. It follows then that simplicity and modesty must always be the keynote of the controlling authority.

A serious defect in the research organization proper is encountered at the very beginning. The Research Institutes are organized on the basis of the particular science, not on recognized branches of farming. The instrument (science) and the subject (agriculture) at once lose contact. The workers in these institutes confine themselves to some aspect of their specialized field; the investigations soon become departmentalized; the steadying influence of first-hand practical experience is the exception rather than the rule. The reports of these Research Institutes describe the activities of large numbers of workers all busy on the periphery of the subject and all intent on learning more and more about less and less. Looked at in the mass, the most striking feature of these institutions is the fragmentation of the subject into minute units. It is true that attempts are made to co-ordinate this effort by such devices as the formation of groups and teams, but as will be shown later (p. 191) this rarely succeeds. Another disquieting feature is the gap between science and practice. It is true that most, if not all, of these establishments possess a farm, but this is mostly taken up with sets of permanent experiments. I know of no research institute in Great Britain besides Aberystwyth where a scientific worker has under his personal control an area of land with his own staff where he can follow the gleam wheresoever it may lead him. Even Aberystwyth stops short before the animal is reached. The improved strains of herbage plants and the method of growing them are not followed to their logical conclusion—a flock of healthy sheep ready for the market and a supply of well-nourished animals by which the breed can be continued.

Has the official machine ever posed to itself such questions as these? What would be the reaction of some Charles Darwin or Louis Pasteur of the future to one or other of these institutes? What would have been their fate if circumstances had compelled them to remain in such an organization, working at some fragment of science? How can the excessive departmentalization of research provide that freedom without which no progress has ever been made in science? Is it rational in such a subject as agriculture to attempt to separate science and practice? Will not the organization of such research always be a contradiction in terms, because the investigator is born not made? The official reply to these questions would form interesting reading.

How does this research organization strike the tillers of the soil for whose benefit it has been created? The farmers complain that the research workers are out of touch with farming needs and conditions; that the results of research are buried in learned periodicals and expressed in unintelligible language; that these papers deal with fragments of the subject chosen haphazard; that the organization of research is so cumbersome that the average farmer cannot obtain a prompt answer to an inquiry and that there are no demonstration farms at which practical solutions of local problems are to be seen.

There seems to be only one effective answer to these objections. The experiment station workers should take their own advice and try out their results. The fruits of this research should be forthcoming on the land itself. All the world over this simple method of publication never fails to secure the respect and attention of the farming community; their response to such messages is always generous and immediate. In Great Britain, however, the retort of the administration takes another line. The idea is fostered that the experiment stations are arsenals of scientific knowledge which actually needs explanation and dilution for the farmer and his land to benefit. Thus in dealing with this point the PEP Report states: 'One of the principal tasks of the administrator is to ensure that the general body of scientific knowledge, including recent results of the research workers' efforts, is brought to the farmer in such a way that he can understand it and apply it on his farm.' The most effective way of doing this is for the organization to demonstrate, in a practical way for all to see, the value of some, at any rate, of these researches. This simple remedy will silence the critics and scoffers; any delay in furnishing it will only add fuel to the fire. After all, a research organization which costs the nation £700,000 a year cannot afford to have its operations called in question by the very men for whose benefit it has been designed. The complaints of the farming community must be removed.

A system not unlike that just described in Great Britain has been adopted in the Empire generally. There is, however, one interesting difference. The official machinery is comparatively simple; the multiplication of agencies and supervisory committees is not so pronounced; the step from the Treasury to the farmer is much shorter. When, however, we come to the research proper, the system is very similar to that which obtains in Great Britain. There is the same tendency to divide research into two groups—fundamental and local; to rely on the piecing together of fragments of science; to extol the advantages of co-operation; to adopt the team rather than the individual. It is the exception rather than the rule to find an investigation in the hands of one competent investigator, provided with land, ample means, and complete freedom.

The completion of an imperial chain of experiment stations for fundamental research was emphasized by a Conference[1] which met in London in 1927. The financial depression, which set in soon after the Report appeared, interfered with this scheme. No additions to the two original links of the chain of five or six super-Research Institutes contemplated—sometimes irreverently referred to as the 'chain of pearls'—have been added to the one in the West Indies (Trinidad) and the other in East Africa (Amani).

Two examples will suffice to illustrate the methods now being employed in this fundamental research work. These are taken from a recent paper by Sir Geoffrey Evans, C.I.E., entitled 'Research and Training in Tropical Agriculture', which appeared in the *Journal of the Royal Society of Arts* of February 10th, 1939. Sir Geoffrey selected the current work on cacao and bananas when explaining how research is conducted at the Imperial College of Tropical Agriculture in Trinidad. He laid great stress on the merits of team work, a method of investigation which we must now examine. These Trinidad examples of research do not stand alone. They resemble what is going on all over the

[1] *Report of the Imperial Agricultural Research Conference*, H.M. Stationery Office, London, 1927.

Empire, including India. Similar work can be collected by the basketful.

In 1930 a study of cacao was commenced in Trinidad in two directions—botanical and chemical. After a preliminary examination of the crop, which is made up of a bewildering number of types, varying widely in fruitfulness and quality, a hundred special trees were selected as a basis for improvement. As cacao does not breed true from seed, methods of vegetative reproduction by means of cuttings and bud wood were first studied. The mechanism of pollination, however, showed that cacao is frequently self-sterile and that many of the special trees required to be cross-pollinated before they could set seed. Suitable pollen parents had then to be found. Manurial experiments on conventional lines led to numerous field experiments all over the island as well as to a detailed soil survey. The biochemical study of the cacao bean produced results described as intricate and baffling; no correlation between the tannin content and quality emerged. The Economics Department of the College investigated the decline of the industry since the War, and established the interesting fact that a cacao plantation reaches its peak in about twenty-five years and then begins to decline. The causes of this decline have been studied and the system of regenerating old plantations by supplying vacancies with high-yielding types has been devised. As, however, the decline of these cacao estates is more likely to be due to worn-out soil than anything else, this method by itself is not likely to succeed. Pests and diseases take their toll of cacao, so the entomologists and mycologists were called in to deal with thrips—the most serious insect pest—and the witch-broom disease—a fungous pest which has done great damage in the West Indies.

The Trinidad investigations on the banana owe their origin to the outbreak of the Panama disease (*Fusarium cubense*) all over the West Indies and the Central American Republics. When the nature of the trouble was established by the mycologists, a search for immune and resistant varieties followed. This included plant breeding, the investigation of the causes of seedlessness, the raising of numerous seedlings, and the search for the ideal parent from which to breed a new commercial banana which is disease-resistant, seedless, of good quality, and capable of standing up to transport conditions. In this work the assistance of the Royal Botanic Gardens at Kew was enlisted; it involved the problem of protecting the banana in the West Indies from disease, including virus, when importing from Malaya (the home of the chief banana of commerce—the Gros Michel) and other places the material needed for the plant-breeding work. The problems of ripening during transport, including a study of the respiration processes during gas storage and the effect of humidity, and the reason for chilling also received attention.

These interesting investigations, which have as their aim the production of higher yields of better cacao and better bananas, have been carried on by what is known as team work. They have necessitated the services of botanists, chemists, mycologists, entomologists and economists, and both have involved considerable expense and much time.

As examples of the way in which the more difficult problems of tropical agriculture are now approached by a number of workers, they are typical of the methods of research everywhere. Many aspects of the cacao and banana problems have been studied; the methods of research have been clearly set out. The workers have evidently spared no pains to achieve success. Nevertheless, the results are negative. The paper under review suggests that matters are still very much in the programme stage; few if any tangible results have been obtained; neither the cacao nor the banana industry has been set on its feet.

If we take a wide view of these two problems and consider: (1) the present methods by which cacao and bananas are grown in the West Indies; (2) the indications furnished by disease that all is not well with these plantations, and (3) the best examples of cacao and banana cultivation to be found in the East where, by means of farm-yard manure only, heavy crops of fine, healthy produce are obtained, the suspicion grows that at least some vital factors have been forgotten in these Trinidad investigations. The spectacular response of cacao trees to humus seems to have been missed altogether and no attention has been given to the significance of the mycorrhizal association in the roots of both cacao trees and bananas. In the cacao and banana plantations in the West Indies, there is a want of balance between the crop and the animal. There

o

is insufficient live stock. There is a disquieting amount of disease and general unthriftiness, which is associated with the absence of conditions suitable for mycorrhizal formation.

Practical experience of the best banana and cacao cultivation in India and Ceylon proves beyond all doubt that the two factors which are essential, if satisfactory yields of high quality are to be obtained and the plantations are to be kept healthy, are: (1) good soil aeration, and (2) regular supplies of freshly prepared humus from animal and vegetable wastes, which are needed to maintain in effective operation the mycorrhizal association. Want of attention to either of these factors is at once followed by loss of quality, by diminished returns, and finally by disease. A better way of dealing with these West Indian problems would have been by good farming methods, including a proper balance between crops and live stock, and by the conversion of all available vegetable and animal wastes into humus.

The Trinidad investigations are quoted as 'an example which can hardly fail to impress the student investigator with the necessity for co-operation'. In reality all they show is how employment can be found for a number of specialists for quite a long time, and indeed what a lot of scientific work can be done by competent workers with purely negative results as far as the yield and the quality of the crop are concerned.

It is not difficult to see the weakness of this method of approach. The problem is never envisaged as a whole and studied in the field from every angle before research on some branch of science is undertaken. Methods of crop improvement are now expected to come from the laboratory and not from the field as they have always done throughout farming history. The control of the team is of necessity very loose. It is normally placed in the hands of persons of administrative rather than practical experience and of limited training in research methods. Often they have other important duties and cannot give the time and thought required. Unable themselves to make a correct diagnosis of the case in the field, their only resource is to go on adding specialist after specialist to their staff in the hope that the study of a fresh fragment of the subject will lead them to some solution. It is almost certain that had the West Indian problems been tackled by one investigator with a real knowledge of farming combined with a wide training in science, and had he been provided with the necessary land, money, and facilities and with complete freedom in conducting the investigation, Sir Geoffrey Evans would have told a very different story. From the point of view of the students at the Trinidad College, it would have been still better to have used these crops for illustrating both methods simultaneously—the banana studied by a single investigator, adequately equipped; cacao by means of a team. In this way the relative merits of the two methods could have been settled for all time. In all probability, two results would have been obtained: (1) the principle that the researcher is the only thing that matters in research would have been established; (2) team work would have ceased to be considered as an effective instrument of investigation.

Team work offers no solution for the evils which result from fragmentation of a research problem. The net woven by the team is often full of holes. Is the fragmentation of the problem accompanied by any other disadvantages? This question is at once answered if we examine any of the major problems of present-day farming. Two British examples will suffice to prove that an inevitable consequence of fragmentation and specialization is loss of direction. Science then loses itself in a maze of detail.

The retreat of the potato crop before blight, eelworm, and virus is one of the most disquieting incidents in British agriculture. One of our most important food crops cannot now be grown successfully on a field scale without a thin film of copper salts; a new rotation of crops from which the potato is omitted until the cysts of the eelworm disappear from the soil; a frequent change of seed from Scotland, Wales, or Northern Ireland. Evidently something is very wrong somewhere, because this crop, when grown in thousands of fertile kitchen gardens throughout the country, is healthy, not diseased. Agricultural science began by fragmenting this potato problem into a number of parts. Potato blight fell within the province of the mycologist; a group of investigators dealt with eelworm; a special experiment station was created for virus disease; the breeding and testing of disease resistant varieties was again a separate branch of the work; the manuring and general agronomy of the crop fell within the province of the agriculturist. The multiplication of workers obscures rather than clarifies this wide biological problem. The fact that these potato diseases exist at all implies that some failure in soil management has occurred. The obvious method of dealing with a collapse of this kind should have been to ascertain the causes of failure rather than to tinker with the consequences of some mistake in management. The net result has been that all this work on the periphery of the subject has not solved the problem of how to grow a healthy potato. This is because direction has been completely lost.

The same story is repeated in manuring: fragmentation has again been followed by loss of direction. Notwithstanding the fact that in the forest Nature has provided examples to copy and in the peat-bog examples to avoid, when devising any rational system of manuring, agricultural science at once proceeded to fragment the subject. For nearly a hundred years some of the ablest workers have devoted themselves to a study of soil nutrients, including trace elements like boron, iron, and cobalt. Green-manuring is a separate subject, so is the preparation of artificial farm-yard manure and the study of the ordinary manure heap. The weight of produce and the cost of manuring overshadow questions of quality. The two subjects which really matter in manuring—the preservation of soil fertility and the quality of the produce—escape attention altogether, mainly because direction has been so largely lost.

The insistence on quantitative results is another of the weaknesses in scientific investigation. It has profoundly influenced agricultural research. In chemistry and physics, for example, accurate records are everything: these subjects lend themselves to exact determinations which can be recorded numerically. But the growing of crops and the raising of live stock belong to biology, a domain where everything is alive and which is poles asunder from chemistry and physics. Many of the things that matter on the land, such as soil fertility, tilth, soil management, the quality of produce, the bloom and health of animals, the general management of live stock, the working relations between master and man, the *esprit de corps* of the farm as a whole, cannot be weighed or measured. Nevertheless their presence is everything: their absence spells failure. Why, therefore, in a subject like this should there be so much insistence on weights and measures and on the statistical interpretation of figures? Are not the means (quantitative results and statistical methods) and the subject investigated (the growth of a crop or the raising of live stock) entirely out of relation the one to the other? Can the operations of agriculture ever be carried out, even on an experiment station, so that the investigator is sure that everything possible has been done for the crop and for the animal? Can a mutually interacting system, like the crop and the soil, for example, dependent on a multitude of factors which are changing from week to week and year to year, ever be made to yield quantitative results which correspond with the precision of mathematics?

The invasion of economics into agricultural research naturally followed the use of quantitative methods. It was an imitation of the successful application of costings to the operations of the factory and the general store. In a factory making nails, for example, it is possible, indeed eminently desirable, to compare the cost of the raw material and the operations of manufacture, including labour, fuel, overhead expenses, wear and tear and so forth, with the output, and to ascertain how and where savings in cost and general speeding up can be achieved. Raw materials, output, and stocks can all be accurately determined. In a very short time a manufacturer with brains and energy will know the cost of every step in the process to the fourth place of decimals. This is because everything is computable. In a similar manner the operations of the general store can be reduced to figures and squared paper. The men in the counting-house can follow the least falling-off in efficiency and in the winning of profit. How very natural it was some thirty years ago to apply these principles to Mother earth and to the farmer! The result has been a deluge of costings and of agricultural economics largely based on guess-work, because the machinery of the soil will always remain a closed book. Mother earth does not keep a pass-book. Almost every operation in agriculture adds or subtracts an unknown quantity to or from the capital of the soil—fertility—another unknown quantity. Any experimental result such as a crop is almost certain to be partly due to the transfer of some of the soil's capital to the profit and loss account of the farmer. The economics of such operations must therefore be based on the purest of guess-work. The results can hardly be worth the paper they are written on. The only things that matter on a farm are these: the credit of the farmer— that is to say what other people, including his labour force and his bank manager, think of him; the total annual expenditure; the total annual income and the annual valuation—the condition of the land and of the live and dead stock at the end of the year. If all these things are satisfactory nothing else matters. If they are not, no amount of costings will avail. Why, therefore, trouble about anything beyond these essentials?

But economics has done a much greater disservice to agriculture than the collection of useless data. Farming has come to be looked at as if it were a factory. Agriculture is regarded as a commercial enterprise; far too much emphasis has been laid on profit. But the purpose of agriculture is quite different from that of a factory. It has to provide food in order that the race may flourish and persist. The best results are obtained if the food is fresh and the soil is fertile. Quality is more important than weight of produce. Farming is therefore a vital matter for the population and ranks with the supply of drinking water, fresh air, and protection from the weather. Our water supplies do not always pay their way; the provision of green belts and open spaces does not yield a profit; our housing schemes are frequently uneconomic. Why, then, should the quality of the food on which still more depends than water, oxygen, or warmth be looked at in a different way? The people must be fed whatever happens. Why not, then, make a supreme effort to see that they are properly fed? Why neglect the very foundation-stone of our efficiency as a nation? The nation's food in the nature of things must always take the first place. The financial system, after all, is but a secondary matter. Economics therefore, in failing to insist on these elementary truths, has been guilty of a grave error of judgement.

In allowing science to be used to wring the last ounce from the soil by new varieties of crops, cheaper and more stimulating manures, deeper and more thorough cultivating machines, hens which lay themselves to death, and cows which perish in an ocean of milk, something more than a want of judgement on the part of the organization is involved. Agricultural research has been misused to make the farmer, not a better producer of food, but a more expert bandit. He has been taught how to profiteer at the expense of posterity—how to transfer capital in the shape of soil fertility and the reserves of his live stock to his profit and loss account. In business such practices end in bankruptcy; in agricultural research they lead to temporary success. All goes well as long as the soil can be made to yield a crop. But soil fertility does not last for ever; eventually the land is worn out; real farming dies.

In the following chapter an example of the type of research needed in the future will be described.

BIBLIOGRAPHY

CARREL, ALEXIS. *Man, the Unknown*, London, 1939.

Constitution and Functions of the Agricultural Research Council, H.M. Stationery Office, London, 1938.

DAMPIER, SIR WILLIAM C. 'Agricultural Research and the Work of the Agricultural Research Council', *Journal of the Farmers' Club*, 1938, p. 55.

EVANS, SIR GEOFFREY. 'Research and Training in Tropical Agriculture'. *Journal of the Royal Society of Arts*, lxxxvii, 1939, p. 332.

LIEBIG, J. *Chemistry in its Applications to Agriculture and Physiology*, London, 1840.

Report of the Imperial Agricultural Research Conference, H.M. Stationery Office, London, 1927.

Report on Agricultural Research in Great Britain, PEP, 16 Queen Anne's Gate, London, 1938.

CHAPTER XIV

A SUCCESSFUL EXAMPLE OF AGRICULTURAL RESEARCH

IN the last chapter the agricultural research of to-day was severely criticized; its many shortcomings were frankly set out; suggestions were made for its gradual amendment. That these strictures are justified will be evident if we examine in detail a piece of successful research carried out on the sugar-cane crop in India during a period of nearly twenty-seven years: 1908–35.

In 1910 the investigations on sugar-cane in Northern India were mainly concentrated in the United Provinces, where a considerable local industry was already in existence. Narrow-leaved, thin canes were planted under irrigation at the beginning of the hot season in March; the crop was crushed by bullock power during the cold weather (January to March); the juice was converted into crude sugar in open pans. The yield was low, a little over one ton to the acre on the average. It was decided to develop this primitive industry and the work, at first purely chemical, was placed in the hands of Mr. George Clarke, the Agricultural Chemist. The choice proved to be a happy one. Clarke combined a first-class knowledge of chemistry and general science with considerable experience of research methods, obtained under Professors Kipping and Pope at Nottingham University College and the School of Technology, Manchester. The son of a south Lincolnshire farmer, he had all his life been familiar with good agricultural practice and had inherited a marked aptitude for farming from a long line of yeomen ancestors. He had therefore acquired the three preliminary qualifications essential for an investigator in agriculture, namely, the makings of a good farmer, a sound training in science, and a first-hand acquaintance with methods of research. It will be seen from what follows that he also possessed the gift of correct diagnosis, the capacity to pose to himself the problems to be investigated, the persistence to solve them, and the drive needed to get the results taken up by the people and firmly welded into the practice of the country-side.

Clarke was most fortunate in the choice of his staff. He had associated with him throughout two Indian officers—S. C. Banerji (afterwards Rai Bahadur) and Sheikh Mahomed Naib Husain (afterwards Khan Bahadur). Banerji, who possessed the dignity and repose of his race, was in charge of the laboratories—always a model of order and efficiency—and was accurate and painstaking almost beyond belief. Naib Husain was of a very different temperament—hot-tempered and full of the energy and drive needed to break new ground in crop production. His absorbing interest in life was the Shahjahanpur farm and the condition of his crops and he never spared himself to make all he undertook a success. Both these men gave their lives to their work and both lived to see their efforts crowned with success, and the Shahjahanpur farm the centre of perhaps the most remarkable example of rural development so far achieved. No European officer in India has ever had more loyal assistants; Indian agriculture has never been served with more devotion. I saw much of their work and watched the growth of the modest but efficient organization they helped to build. It is a great regret to me that they are not here to read this genuine tribute of admiration and respect from a fellow worker of another race.

It had been the custom till 1912 in the United Provinces to keep their scientific officers isolated from the practical side of agricultural improvement, and there were no clear ideas how the combined scientific and practical problems connected with cane cultivation should be tackled. It did not strike any one that it would be necessary for the scientific investigator to grow the crop and master the local agriculture before any improvement could be devised and tried. When therefore in 1911 Clarke asked to be provided with a farm, there was a good deal of discussion and some amazement. The matter was referred to the All-India Board of Agriculture in 1911, where the proposal was severely criticized. The agricultural members did not like the idea of scientific men having land of their own; the representatives of science considered they would lose caste if another of their number took up farming.

In 1912 I happened to be on tour in the United Provinces when the matter came up for final decision. The Director of Agriculture asked my advice. I strongly supported Clarke's proposal and assured the authorities that great things would result if they gave their agricultural chemist the best farm they could, and then left him alone to work out his own salvation. This carried the day; a special sugar-cane farm was established in 1912 near Shahjahanpur on the bank of the Kanout river and on one of the main roads leading into the town. From 1912 to 1931, an unbroken period of nineteen years, Clarke remained in charge of Shahjahanpur in addition to the three posts he held: Agricultural Chemist (1907-21), Principal of the Agricultural College (1919-21), and Director of Agriculture (1921-31). From 1912 to 1921 he was there nearly every week-end. Until he became Director of Agriculture in 1921, he was at Shahjahanpur every year from Christmas until March for the sugar-cane harvest and the planting of the next year's crop. It was during these periods that he gained a first-hand knowledge of the Indian village, its people, its fields, and its agricultural problems, which was to stand him in such good stead when he was called upon to direct the agricultural development of the Provinces in the early years of the Montagu-Chelmsford Reforms.

The sugar-growing tract of northern India, the most important in the country, is a broad strip of deep alluvial land about 500 miles in length skirting the Himalayas. It begins in Bihar; it ends in the Punjab, and reaches its greatest development in the Revenue Divisions of Gorakhpur, Meerut, and Rohilkhand of the United Provinces. The soil is easily cultivated and is particularly suitable for the root development of the sugar-cane. The climate, however, is not particularly favourable as the growing period is so short and confined to the rainy season—the last half of June. July, August, and September—when the moist tropical conditions created by the south-west monsoon are established. The rains are followed by the cold season (October 15th to March 15th) during which very little rain is received. After the middle of March the weather again changes, becoming very hot and dry till the break of the rains in June. During the hot weather the cane, which is usually planted towards the end of February, has to be irrigated.

When work was started in 1912, the yield of stripped cane on 95 per cent. of the sugar-cane area of the United Provinces was only 13 tons to the acre, producing just over 1 ton of crude sugar (*gur*). The land was fallowed during the previous rains and was well prepared for the crop by 15 to 20 shallow ploughings with the native plough. As in many other Indian crops a good balance has been established, as a result of age-long experience, between the methods of cultivation and the economic capacity of the indigenous varieties. The methods of cultivation, the nitrogen supply, and the kinds grown were all in correct relation the one to the other. These varieties had been cultivated for at least twenty centuries and were thin, short, and very fibrous with juice rich in sugar in good seasons. They remind one more of the wild species of the genus *Saccharum* than of the thick sugar-canes found in tropical countries. Five or six varieties are grown together, each with a name, usually of Sanskrit origin, denoting their qualities, and each is readily recognized by the people.

The types of cane grown by the cultivators in the Rohilkhand Division were first separated and an attempt was made to intensify the cultivation of the best. The yield of cane was raised from 13 to 16 tons per acre without deterioration in the quality of the juice, but further intensification did not succeed. As much as 27 tons to the acre was obtained, but the thin, watery juice contained so little sugar that it was not worth extracting. The variety and the improved soil conditions were not in correct relation. The leaf area developed by the indigenous varieties was insufficient, in the short growing period of North India, to manufacture the cellulose required for the fibre and other tissue of so large a crop, and enough sugar to make the juice of economic value. This was a most important experiment as it defined the general problem and showed definitely what had to be done. To raise the out-turn of sugar in the United Provinces, a combination of intensive methods of cultivation with more efficient varieties, adapted to the very special climatic conditions, would be necessary. These two problems were taken up simultaneously: in all the subsequent work the greatest care was taken to avoid the fragmentation of the factors, a rock on which so much of the agricultural research of the present day founders.

Attention was then paid to the two chief factors underlying the problem: (1) the discovery of a suitable cane, and (2) the study of intensive cane growing with the object of finding out the maximum yield that could be obtained.

The collection of cane varieties at Shahjahanpur included a Java seedling—POJ 213—which was exactly suited to the local soil and climate and which responded to intensive cultivation. This Java cane was a hybrid. Its pollen parent was the Rohilk-hand variety *Chunni*, which had been given to the Dutch experts, who visited India twenty years earlier, by the Rosa Sugar Factory. *Chunni* was immune to *sereh*—a serious disease then threatening the sugar industry in Java—and, when crossed with rich tropical canes, produced immune or very resistant seedlings of good quality, widely known throughout the world as POJ (Passoerean Ost-Java) seedlings. POJ 213 proved invaluable during the early stages of the Shahjahanpur work. It was readily accepted by the culti-vators among whom it was known as 'Java'. A large area was soon grown in Rohilkhand and it saved the local sugar industry then on the verge of extinction. Most important of all it created an interest in new kinds of sugar-cane and prepared the ground for the great advance, which came a few years later when a Coim-batore seedling—Co 213—raised by the late Dr. Barber, C.I.E.—replaced it.

Clarke had not been long in the United Provinces before he noticed that the soil of the Gangetic plain could be handled in much the same way as that of the Holland Division of Lincoln-shire, where the intensive cultivation of potatoes had been intro-duced sixty or seventy years before and brought to a high state of perfection. Both soils are alluvial though of widely different date. The problems connected with the intensive cultivation of potatoes in Lincolnshire and sugar-cane in the United Provinces have a great deal in common. Both crops are propagated vegetatively, and in both it is a most important point to produce the soil condi-tions necessary to develop the young plant quickly, so that it is ready to manufacture and store a large quantity of carbohydrate in a short period of favourable climatic conditions. When the intensification of sugar-cane cultivation was begun at Shah-jahanpur, the lessons learnt in the potato fields of Lincolnshire were at once applied. During the fallow which preceded the crop, the land was cultivated and, as soon as possible after the retreat of the monsoon, the farm-yard manure was put on and ploughed in. This gave the time necessary for a valuable supply of humus to be formed in the top layer of the soil. The sugar-cane was planted in shallow trenches 2 feet wide, 4 feet from centre to centre. The soil from each trench was removed to a depth of 6 inches and piled on the 2 feet space left between each two trenches, the whole making a series of ridges as illustrated in Fig. 6.

As soon as the trenches are made in November, they are dug with a local tool (*kasi*) to a further depth of 9 inches, and the

FIG. 6. Trench system at Shahjahanpur

oilcake, or whatever concentrated organic manure is available, is thoroughly mixed with the soil of the floor of the trenches and allowed to remain, with occasional digging, till planting time in February. The thorough cultivation and manuring of the trenches at least two months, and preferably three, before the canes are planted, proved to be essential if the best results were to be obtained. Readers familiar with the methods of sugar-cane culti-vation in Java will at once realize that this Shahjahanpur method is a definite advance on that in use in Java, in that the use of artificials is entirely unnecessary. Hand-made trenches always give better yields than those made by mechanical means—an interesting result, which has often been obtained elsewhere, but which has never been adequately explained. It may be that speed in cultivation is an adverse factor in the production of tilth.

At first heavy dressings of organic manures like castor cake meal at the rate of about 2,870 lb. to the acre were used in the trenches. This contains about 4·5 per cent. of nitrogen, so that 2,870 lb. is equivalent to 130 lb. of nitrogen to the acre. This heavy manuring, however, was soon reduced after the introduction of the method of green-manuring described below. When green-manuring was properly carried out, the dung applied before making the trenches could be reduced to half or even less.

At first the trenches were irrigated about a month before planting and lightly cultivated when dry enough. These operations promoted the decay of the manure, and allowed for abundant soil aeration. The canes were planted in the freshly dug moist rich earth towards the end of February. Later this preliminary watering was dispensed with. The cuttings were planted in the dry earth, and lightly watered the next day. This saved one

FIG. 7. Earthing up sugar-cane at Shahjahanpur, July 10th, 1919

irrigation and proved to be an effective protection from white ants (*Termites*), which often attacked the cane cuttings unless these started growth at once and the young plants quickly established themselves. Four light waterings, followed in every case by surface cultivation, were necessary before the break of the monsoon in June. As soon as the young canes were about 2 feet high, and were tillering vigorously, the trenches were gradually filled in, beginning about the middle of May and completing the operation by the end of the month. Before the rains began, the earthing up of the canes commenced. It was completed by about the middle of July (Fig. 7).

One of the consequences of earthing up canes, grown in fertile soil, observed by Clarke was the copious development of fungi which were plainly visible as threads of white mycelium all through the soil of the ridges, and particularly round the active roots. As the sugar-cane is a mycorrhiza-former there is little doubt that the mycelium, observed in such quantities, was connected with the mycorrhizal association. The provision of all the factors needed for this association—humus, aeration, moisture, and a constant supply of active roots—probably explains why such good results have always followed this method of growing the cane and why the crops are so healthy. When grown on the flat, want of soil aeration would always be a limiting factor in the full establishment of the mycorrhiza.

The operation of earthing up serves four purposes: (1) the succession of new roots, arising from the lower nodes, thoroughly combs the highly aerated and fertile soil of the ridges; (2) the conditions suitable for the development of the mycorrhizal association are provided; (3) the standing power of the canes during the rains is vastly improved; and (4) the excessive development of colloids in the surface soil is prevented. If this earthing up is omitted, a heavy crop of cane is almost always levelled by the monsoon gales; crops which fall down during the rains never give the much-prized light coloured crude sugar. The production of colloids in the surface soil, when the canes are grown on the flat, always interferes with soil aeration during the period when sugar is being formed; crops which ripen under conditions of poor soil aeration never give the maximum yield.

An essential factor in obtaining the highest efficiency in this ridging system is good surface drainage. This was achieved by lowering the earth roads and paths which, when grassed over, acted as very efficient drains for carrying off the excess rainfall during the monsoon. The surface water collected in the trenches which were suitably connected with the system of lowered paths and roads. By this means the drainage crept away, in thin sheets of clear water, to the river without any loss of organic matter or of fine soil particles. The grass carpet acted as a most efficient filter and was at the same time manured. The roads yielded good crops of grass for the work cattle. This simple device should be utilized wherever possible to prevent both water-logging and soil erosion.

The results of this intensive method of cane cultivation—based on the growth of efficient varieties, proper soil aeration, good surface drainage, carefully controlled irrigation, and an adequate supply of organic matter—were astounding. In place of 13 tons of cane and just over a ton of sugar per acre, a yield of just over or just under 36 tons of cane and 3½ tons of sugar per acre was obtained for a period of twenty years—year in and year out. These

are the figures for the farm as a whole. The yield of sugar had
been trebled. Such a result has rarely been obtained for any
crop in so short a time and by such simple means. In a few cases
yields as high as 44 tons of cane and 4½ tons of sugar were obtained,
figures which probably represent the highest possible production
in the climate of the United Provinces.

In working out this method of cane growing in northern India,
two critical periods in sugar production were observed: (1) May
and early June when the tillers and root system are developing,
and (2) August and September when the main storage of sugar
takes place. A check received at either of these periods permanently reduces the yield. The acre yield of sugar is positively
and closely correlated with the amount of nitrate nitrogen in the
soil during the first period and with soil moisture, soil aeration, and
atmospheric humidity in the second period. Any improvement in
cane growing must therefore take into account these two principles.

A new method of intensive cane cultivation had been devised
and put into successful practice on a field scale at a State Experiment Station; the first step in improving sugar production had
been taken. It was now necessary to fit this advance into a system
of agriculture made up of a multitude of small holdings, varying
from an average of 4 acres in the eastern districts to 8 acres in
the western half of the province. Each holding is not only minute
but it is divided into tiny fields, scattered over the village area,
which is by no means uniform in fertility. Moreover, the farmers
of these small holdings possess practically no capital for investment in intensive agriculture. How could the average cultivator
obtain the necessary manure? The solution of this problem
entailed a detailed study of the nitrogen cycle on the Gangetic
alluvium, the relation between climate, methods of cultivation,
and the accumulation of soil nitrate as well as the discovery of
what amounts to a new method of green-manuring for sugar-cane.
These investigations were set in motion the moment the possibilities of intensive cane growing became apparent.

The study of the nitrogen cycle, in any locality, naturally
includes an intimate acquaintance with the local agriculture.
The outstanding feature of the agricultural year in the United
Provinces is the rapidity with which the seasons change, and the

PLATE VII

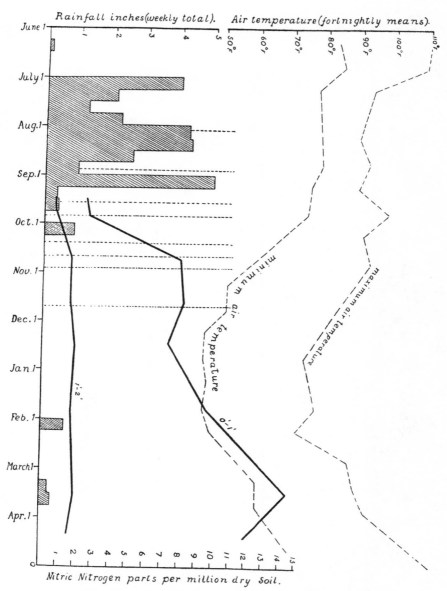

Rainfall inches (weekly total).   Air temperature (fortnightly means).

Nitric Nitrogen parts per million dry Soil.

NITRATE ACCUMULATION IN THE GANGETIC ALLUVIUM

wide variation in their character. The most important of these abrupt transitions are: (1) the change from the excessive dryness and high temperatures of April, May, and early June to the moist tropical conditions which set in when the summer crops are sown at the end of June, immediately after the break of the south-west monsoon, and (2) the sudden transition from high humidity, high temperature, and a saturated condition of the soil, at the end of the monsoon in September, to the dry temperate conditions which obtain when the autumn sowings of food crops take place in October. These sudden seasonal changes impose definite limits on what can be done to increase production. There is very little time for the preparation of the land, or for the manufacture of plant food by biological agencies; the period of active growth of the crop is severely limited. The former influences the methods of cultivation and manuring; the latter the selection of varieties. The wide difference between the two agricultural seasons in the United Provinces is best realized in the autumn (November and December), when crops of ripening cane and growing wheat are to be seen side by side in adjacent fields.

How do the summer and autumn crops, often raised in this area under extensive methods, manage to obtain a supply of nitrate without any added manure, and how is it that the soil fertility of the Gangetic alluvium remains so constant? To begin to answer these questions, soil borings from a typical unmanured area—fallowed after the wheat crop which was removed in April—were systematically examined and the nitric nitrogen estimated directly by Schloesing's method.

The results, as well as the details of temperature and rainfall, are given in Plate VII. The curve brings out clearly: (1) the large and rapid formation of nitrate as the temperature rises in February and March, just at the time when the young sugar-cane plant takes in its supply of nitrogen, (2) the almost complete disappearance of nitrates from the soil after the first falls of heavy rain—these are partly washed out and partly immobilized by fungous growth, provided the humus content of the soil has been maintained, (3) the absence of nitrification in the saturated soil during the monsoon, (4) another accumulation of nitrate (less rapid and less in quantity than in the spring) which occurs in the autumn,

following the drying of the soil at the end of the rains, combined
with improved aeration, the result of frequent surface cultivation.
Five ploughings to a depth of 3 inches, followed by the levelling
beam, were given between September 25th and November 30th.
These accumulations, obviously the result of biological processes,
fit in with the quantitative requirements of the summer and
autumn crops, which need immediate supplies of nitrogen as soon
as the seed germinates.

When we compare these results on nitrate accumulation with
what the Indian cultivator is doing, we are lost in admiration of
the way he sets about his task. With no help from science, and by
observation alone, he has in the course of ages adjusted his
methods of agriculture to the conservation of soil fertility in a
most remarkable manner. He is by no means the ignorant and
backward villager he is sometimes represented to be, but among
the most economical farmers in the world as far as the management of the potent element of fertility—combined nitrogen—goes,
and tropical agriculture all over the world has much to learn from
him. The sugar grower of the great plains of India cannot take
a heavy overdraft of nitrogen from his soil. He has only a limited
store—the small current account provided by non-symbiotic
nitrogen fixation and the capital stock of humus needed to maintain the crumb structure and the general life of the soil. He must
make the most of his current account; he dare not utilize any of
his capital. He has in the course of ages instinctively devised
methods of management which fulfil these conditions. He does
not over-cultivate or cultivate at the wrong time. Nothing is done
to over-oxidize his precious floating nitrogen or to destroy his
capital of humus. He probably does more with a little nitrogen
than any farmer in the world outside China. For countless ages
he has been able to maintain the present standard of fertility.

If the production of sugar was to be raised, obviously the first
step was to provide more nitrate for the critical growth period of
May and June, when the tillers and root system are developing.
The conventional method would be to stimulate the crop by the
addition of factory-made and imported fertilizers such as sulphate
of ammonia. There are weighty objections to such a course. The
cultivator could not afford them; the supply might be cut off in

time of war; the effect of adding these substances to the soil would be to upset the balance of soil fertility—the foundation of the Indian Empire—by setting in motion oxidation processes which would eat into India's capital, and burn up the vital store of soil humus. Increased crops would indeed be obtained for a few years, but at what a cost—lowered soil fertility, lowered production, inferior quality, diseases of crops, of animals, and of the population, and finally diseases of the soil itself, such as soil erosion and a desert of alkali land! To place in the hands of the cultivator such a means of temporarily increasing his crops would be more than a mere error of judgement: it would be a crime. The use of artificials being ruled out altogether, some alternative source of nitrogen had to be found.

Any intensive method of sugar growing in the United Provinces must accomplish two things: (1) the normal accumulations of nitrates in the soil at the beginning of the rains must be fully utilized, and (2) the content of soil organic matter must be raised and the biological processes speeded up so that these natural accumulations can be increased.

The problem of making use of the nitrate naturally formed in order to raise the content of organic matter was solved in a very neat fashion—by a new method of green-manuring. The fallow, which ordinarily precedes cane, was used to grow a crop of *san* hemp with the help of about 4 tons of farm-yard manure to the acre. This small dressing of cattle manure had a remarkable effect on the speed of growth and also on the way the green crop decayed when it was ploughed in. Yields of 8 tons of green-manure were produced in about 60 days which added nearly 2 tons of organic matter, or 75 lb. of nitrogen, to each acre. In this way the nitrates accumulated at the break of the rains were absorbed and immobilized; a large mass of crude organic material was provided by the green manure itself and by the small dressing of farm-yard manure applied before sowing. Sheet composting took place in the surface soil.

The early stages of decomposition need ample moisture. The rainfall after the green crop was ploughed in was carefully watched. If it was less than 5 inches in the first fortnight in September, the fields were irrigated. In this way an abundant fungous growth was secured on the green-manure as the land slowly dried. The conversion of the whole of the green crop into humus was not complete until the end of November. Nitrification then began, slowly, owing to the low temperatures of the cold weather in North India and at a season when there is little risk of loss. It was not until the newly planted canes were watered at the end of February and the temperature rose at the beginning of the hot weather that all the available nitrogen in the freshly prepared humus was rapidly nitrified to meet the growing needs of the developing root system of the cane. This means simply that a definite time is required for the formation of humus, whether it takes place in the soil by sheet composting or in the compost heap, a longer period being required in the soil than in the heap. The gradual filling of the trenches and the watering of the canes during the hot weather continued the nitrification process, which was carried a stage farther by the earthing up of the canes. The provision of drainage trenches between the rows reduced to a minimum any losses of nitrogen due to poor soil aeration following the formation of soil colloids. The canes were thus provided with ample nitrate throughout the growth period. The conditions necessary for the mycorrhizal association were also established.

The effects of green-manuring on the nitrate supply is shown in Plate VIII. It will be seen that the enrichment of the soil with humus has markedly increased the amount of nitrate formed during the crucial period (March to June) when the rapidly growing cane absorbs most of its supplies.

The yields of cane and of raw sugar of twenty-seven randomized plots in the green-manured and control plots are given below:

TABLE II

Effect of green manuring on sugar-cane

	Sugar-cane maunds (82⅔ lb.), per acre	Raw sugar maunds, per acre	Dry matter maunds, per acre
Green-manure . .	847·0±32·0	87·0±3·6	246·0±8·0
Control . . .	649·0±22·0	67·2±2·6	200·1±6·6

These control plots are representative of the fertile plots of the Shahjahanpur Experiment Station, not of the fields of the cultivators.

PLATE VIII

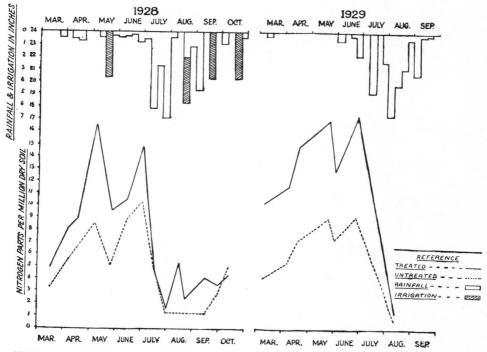

NITRATE ACCUMULATION, GREEN-MANURE EXPERIMENT, SHAHJAHANPUR,
1928–9

The crop in the field is shown in Plate IX. The practical result of this simple method of intensive cane growing worked out at a profit of £6 an acre. These satisfactory results were reflected in the annual statement of income and expenditure of the Shahjahanpur Experiment Station. For many years income exceeded expenditure by about 50 per cent.

The question therefore of the practical value of the work done at this Station needed no argument. It was obvious. By the help of green-manure alone, supplemented by a small dressing of cattle dung, the yield of cane was raised from 13 to over 30 tons to the acre; the yield of sugar from 1 ton to over 3 tons.

The effect of the intensive cultivation of sugar-cane is not confined to that crop. The residual fertility and the deep cultivation of the trenches enabled bumper crops of Pusa wheat and gram —the two rotation crops grown with cane at Shahjahanpur—to be obtained. These were more than three times the average yields obtained by the cultivators. In one case in a field of $3\frac{1}{2}$ acres, Pusa 12 wheat gave 35 maunds to the acre with one irrigation of 4 inches in November. The preceding cane crop was Ashy Mauritius, which yielded 34·7 tons to the acre.

In the early days of the Shahjahanpur farm the effect of the cane trenches on the following wheat crop was very pronounced. The surface of the wheat fields resembled a sheet of corrugated iron, the ridges corresponding to the trenches. After a few sugar crops this condition passed off, and the wheat appeared uniform; all the land had been raised to the new level of fertility.

It is now possible to record the various stages passed through in this study of intensive sugar production:

1. The unimproved crop in an average year yielded 350 maunds per acre (1 maund = $82\frac{2}{7}$ lb.; 27·2 maunds = 1 ton).

2. The best indigenous varieties, grown at their maximum capacity with slightly deeper ploughing than the cultivator gives and a small quantity of manure, gave 450 maunds per acre.

3. The introduction of new varieties like POJ 213 and Co 213, grown on the flat with the same cultivation as (2) above, gave 600 maunds per acre. The increment due to variety was therefore 150 maunds per acre (600—450).

4. The use of new varieties plus green-manure on the flat without trenches yielded 800 maunds per acre. The increment due to green-manure was therefore 200 maunds per acre (800—600).

5. Intensive cultivation in trenches, with green-manure plus the manuring of the trenches with castor cake meal at the rate of 1,640 lb. per acre, yielded 1,000 maunds per acre. The increment due to improved soil aeration and an adequate supply of humus in the trenches was therefore 200 maunds per acre (1,000—800).

6. The very highest yield ever obtained at Shahjahanpur by the use of (5) above was 1,200 maunds per acre. The additional increment, when all the factors were functioning at or near their optima was 200 maunds per acre (1,200—1,000).

It will be seen that a combination of variety, green manuring, and correct soil management, including the manuring of the trenches, added 650 maunds per acre (1,000—350) and that increases of 850 maunds per acre are possible (1,200—350). In the intensive trench method the plant attains an exceptionally high efficiency in the synthesis of carbohydrates. In an out-turn test at Shahjahanpur 1,200 maunds of stripped cane per acre contained 17 per cent. of fibre (mostly pure cellulose), 12 per cent. of sucrose, and 1 per cent. of invert sugar. The total quantity of carbohydrate synthesized per acre in about four months of active growth was 204 maunds of cellulose, 144 maunds of sucrose, and 12 maunds of invert sugar; in all 360 maunds (13.2 tons) per acre of carbohydrates. This means 3·3 tons per acre during the period of active growth when every assistance as regards choice of variety, soil fertility, and soil management had been provided.

We have in these Shahjahanpur results a perfect example of the manufacture of humus by means of a green-manure crop and its utilization afterwards. Success depends on two things: (1) a knowledge of the nitrogen cycle and of the conditions under which humus is manufactured and utilized, and (2) an effective agricultural technique based on these biological principles.

The stage was now set for getting the Shahjahanpur results taken up by the cultivators. Two questions had to be settled:

1. Should an attempt be made to introduce the full Shah-jahanpur methods—improved green-manuring, manured trenches, and new varieties—or should a beginning be made with green-manuring plus new varieties with or without trenches according to circumstances? It was finally decided to introduce the new variety along with the new method of green-manuring and to omit the trenches. This decision was made because of the scarcity of manure for the full Shahjahanpur method. This difficulty, however, was removed in 1931, the year Clarke left India, with the introduction of the Indore Process, which could have provided every village with an adequate supply of manure for intensive methods.

2. Should a large organization be created for bringing the results to the notice of the villagers? At this point the Agricultural Department was transferred to the control of an Indian Minister, and Clarke became Director of Agriculture in the United Provinces and a member of the Legislative Council, offices which he held with short temporary breaks for ten years. He was thus provided with administrative powers for developing to the full the results of his work as a scientific investigator. The first two Ministers under whom he served, Mr. C. Y. Chintamani (now Sir Chirravoori) and the Nawab of Chhatari, though of widely different political views, were in complete agreement regarding the necessity of agricultural development and both gave their unqualified support to the proposals which were placed before them, while in the Council itself members of every shade of opinion, from the extreme Left to the extreme Right, demanded a large extension of agricultural work. The annual debate on the agricultural budget was one of the events of the year. From 1921 to 1931 the financial proposals of Government for agricultural improvement were passed without an adverse vote of any kind. This was incidentally an example of the success that can be obtained under a popular Government in India by a Department when it is backed by efficient technical work. A large number of Members of the Council were influential landlords, deeply interested in the development of the country-side, and they, and many others outside the Council, were anxious to take up something of practical value. It was therefore decided to start, in the main sugar-growing

areas, State-aided private farms for the purpose of demonstrating the new methods of growing cane and providing the large quantity of planting material required for the extension of the area under the new varieties.

The amount of aid given by the State was small, about Rs. 2,000 to Rs. 3,000 for every farm. An agreement was entered into, between the landlords and the Agricultural Department, by which the former agreed to put down a certain area under Co 213, to green-manure the land and to conduct the cultivation on Shah-jahanpur lines. Cuttings were to be supplied to the locality at a certain fixed rate. In this way the example and influence of the landlords was secured at very small cost. At the same time the hold of the Agricultural Department on the country-side was increased and strengthened—the landlords became to all intents and purposes an essential portion of the higher staff of the Agricultural Department. They differed, however, from the ordinary District staff in two important respects: (1) they possessed an influence far surpassing that of the ablest members of the Agricultural Department; (2) they were unofficial and unpaid.

The contribution of the Indian landowners to this work was of the greatest importance. Without their active support and their public spirit in opening farms, and thus setting up a multitude of local centres for demonstrating on a practical scale the new method of sugar production, and at the same time providing the material for planting at a low rate, the Agricultural Department would have had to fall back on itself for getting the improvements taken up. In place of the demonstration farms, provided by the landlords at an exceedingly low cost, the Government would have had to acquire land and start local farms for advertising the new method and for the supply of plant material. The cost would have been colossal and quite beyond the resources of Government. In place of the influence and personal interest of the natural leaders of the country-side, the Agricultural Department would have had to rely on the work of a host of low-paid subordinates, and would have had to increase its inspecting staff out of all knowledge. An unwieldy and expensive organization would have been the result. All this was rendered unnecessary by the admirable system of private farms. The idea of using the landlords in agriculture

PLATE IX

Treated

Control

GREEN-MANURE EXPERIMENT, SHAHJAHANPUR, 1928-9

began in Oudh in 1914, when a number of private farms were started on the estates of the Talukdars for demonstrating the value of the new Pusa wheats and for producing the large quantities of seed needed. Clarke extended this idea to all parts of the Provinces, and showed how the results obtained at one experimental farm could be expanded rapidly and effectively by enlisting the active help of the landowners. He provided them with an opportunity, eagerly taken up, of showing their value to the community—that of leadership in developing the country-side by practical examples of better agriculture which their tenants and neighbours could copy. *Landlords all the world over will act in a similar way once the Agricultural Departments can provide them with results of real value.*

The magnitude of these operations and the speed with which they were conducted will be obvious from the following summary of the final results. In 1916–17 Clarke received about 20 lb. of cuttings of Co 213 from Coimbatore for trial. In 1934–5, 33,000,000 tons of this variety were produced in the United Provinces. The value of the crop of Co 213 cane to the cultivator, at the low minimum rate fixed by Government for sugar-cane under the Sugar-Cane Act of 1934, was over £20,000,000, more than half of which was entirely new wealth. The value of the sugar which could be manufactured from this was £42,000,000. A large sum was distributed in factory wages, salaries, and dividends, to say nothing of the benefit to the British engineering trade and the effect on employment in Great Britain of orders for over £10,000,000 worth of machinery for new factories. There was no question of glutting an over-stocked market in India. All the additional sugar and sugar products were readily absorbed by the local markets.

Here we have a successful effort in directed economy—the development of imperial resources by simple technical improvements in agriculture, rendered possible by the protection of a valuable imperial market by a straight tariff.

Since Clarke retired from the Indian Agricultural Department, two new factors—both favourable—have been in operation, which make it easy to follow up this initial success. The difficulty of producing enough humus in the villages for the trench system has been removed by the Indore system of composting. Irrigation has been improved in the sugar-growing tracts by the completion of the Sarda Canal and the provision of cheap electric power for raising water from wells. The two essentials for intensive agriculture—humus and water—are now available. Before long a detailed account of the progress made by the introduction of the full Shahjahanpur method will no doubt be available. The story begun in this chapter can then be carried on another stage. It will make interesting and stimulating reading.

BIBLIOGRAPHY

CLARKE, G., BANERJEE, S. C., NAIB HUSAIN, M., and QAYUM, A. 'Nitrate Fluctuations in the Gangetic Alluvium, and Some Aspects of the Nitrogen Problem in India', *Agricultural Journal of India*, xvii, 1922, p. 463.

CLARKE, G. 'Some Aspects of Soil Improvement in Relation to Crop Production', *Proc. of the Seventeenth Indian Science Congress*, Asiatic Society of Bengal, Calcutta, 1930, p. 23.

PART V
CONCLUSIONS AND SUGGESTIONS
CHAPTER XV
A FINAL SURVEY

THE capital of the nations which is real, permanent, and independent of everything except a market for the products of farming, is the soil. To utilize and also to safeguard this important possession the maintenance of fertility is essential.

In the consideration of soil fertility many things besides agriculture proper are involved—finance, industry, public health, the efficiency of the population, and the future of civilization. In this book an attempt has been made to deal with the soil in its wider aspects, while devoting due attention to the technical side of the subject.

The Industrial Revolution, by creating a new hunger—that of the machine—and a vast increase in the urban population, has encroached seriously on the world's store of fertility. A rapid transfer of the soil's capital is taking place. This expansion in manufacture and in population would have made little or no difference had the waste products of the factory and the town been faithfully returned to the land. But this has not been done. Instead, the first principle of agriculture has been disregarded: growth has been speeded up, but nothing has been done to accelerate decay. Farming has become unbalanced. The gap between the two halves of the wheel of life has been left unbridged, or it has been filled by a substitute in the shape of artificial manures. The soils of the world are either being worn out and left in ruins, or are being slowly poisoned. All over the world our capital is being squandered. The restoration and maintenance of soil fertility has become a universal problem.

The outward and visible sign of the destruction of soil is the speed at which the menace of soil erosion is growing. The transfer of capital, in the shape of soil fertility, to the profit and loss account of agriculture is being followed by the bankruptcy of the land. The only way this destructive process can be arrested is by restoring the fertility of each field of the catchment area of the rivers which are afflicted by this disease of civilization. This formidable task is going to put some of our oversea administrations to a very severe test.

The slow poisoning of the life of the soil by artificial manures is one of the greatest calamities which has befallen agriculture and mankind. The responsibility for this disaster must be shared equally by the disciples of Liebig and by the economic system under which we are living. The experiments of the Broadbalk field showed that increased crops could be obtained by the skilful use of chemicals. Industry at once manufactured these manures and organized their sale.

The flooding of the English market with cheap food, grown anywhere and anyhow, forced the farmers of this country to throw to the winds the old and well-tried principles of mixed farming, and to save themselves from bankruptcy by reducing the cost of production. But this temporary salvation was paid for by loss of fertility. Mother earth has recorded her disapproval by the steady growth of disease in crops, animals, and mankind. The spraying machine was called in to protect the plant; vaccines and serums the animal; in the last resort the afflicted live stock are slaughtered and burnt. This policy is failing before our eyes. The population, fed on improperly grown food, has to be bolstered up by an expensive system of patent medicines, panel doctors, dispensaries, hospitals, and convalescent homes. A C3 population is being created.

The situation can only be saved by the community as a whole. The first step is to convince it of the danger and to show the road out of this impasse. The connexion which exists between a fertile soil and healthy crops, healthy animals and, last but not least, healthy human beings must be made known far and wide. As many resident communities as possible, with sufficient land of their own to produce their vegetables, fruit, milk and milk products, cereals, and meat, must be persuaded to feed themselves and to demonstrate the results of fresh food raised on fertile soil. An important item in education, both in the home and in the school, must be the knowledge of the superiority in taste, quality, and keeping power of food, like vegetables and fruit, grown with humus, over produce raised on artificials. The women of England—the mothers of the generations of the future—will then exert their influence in food reform. Foodstuffs will have to be graded, marketed, and retailed according to the way the soil is manured. The urban communities (which in the past have prospered at the expense of the soil) will have to join forces with rural England (which has suffered from exploitation) in making possible the restitution to the country-side of its manurial rights. All connected with the soil—owners, farmers, and labourers—must be assisted financially to restore the lost fertility. Steps must then be taken to safeguard the land of the Empire from the operations of finance. This is essential because our greatest possession is ourselves and because a prosperous and contented country-side is the strongest possible support for the safeguarding of the country's future. Failure to work out a compromise between the needs of the people and of finance can only end in the ruin of both. The mistakes of ancient Rome must be avoided.

One of the agencies which can assist the land to come into its own is agricultural research. A new type of investigator is needed. The research work of the future must be placed in the hands of a few men and women, who have been brought up on the land, who have received a first-class scientific education, and who have inherited a special aptitude for practical farming. They must combine in each one of them practice and science. Travel must be included in their training because a country like Great Britain, for instance, for reasons of climate and geology, cannot provide examples of the dramatic way in which the growth factors operate.

The approach to the problems of farming must be made from the field, not from the laboratory. The discovery of the things that matter is three-quarters of the battle. In this the observant farmer and labourer, who have spent their lives in close contact with Nature, can be of the greatest help to the investigator. The views of the peasantry in all countries are worthy of respect; there is always good reason for their practices; in matters like the culti-vation of mixed crops they themselves are still the pioneers. Association with the farmer and the labourer will help research to abandon all false notions of prestige; all ideas of bolstering up their position by methods far too reminiscent of the esoteric priesthoods of the past. All engaged on the land must be brother cultivators together; the investigator of the future will only differ from the farmer in the possession of an extra implement—science —and in the wider experience which travel confers. The future standing of the research worker will depend on success: on ability to show how good farming can be made still better. The illusion that the agricultural community will not adopt improvements will disappear, once the improver can write his message on the land itself instead of in the transactions of the learned societies. The natural leaders of the country-side, as has been abundantly proved in rural India, are only too ready to assist in this work as soon as they are provided with real results. No special organization, for bringing the results of the experiment stations to the farmer, is necessary.

The administration of agricultural research must be reformed. The vast, top-heavy, complicated, and expensive structure, which has grown up by accretion in the British Empire, must be swept away. The time-consuming and ineffective committee must be abolished. The vast volume of print must be curtailed. The expenditure must be reduced. The dictum of Carrel that 'the best way to increase the intelligence of scientists would be to reduce their number' must be implemented. The research applied to agriculture must be of the very best. The men and women who are capable of conducting it need no assistance from the administration beyond the means for their work and protection from interference. One of the chief duties of the Government will be to prevent the research workers themselves from creating an organization which will act as a bar to progress.

The base line of the investigations of the future must be a fertile soil. The land must be got into good heart to begin with. The response of the crop and the animal to improved soil conditions must be carefully observed. These are our greatest and most profound experts. We must watch them at work: we must pose to them simple questions; we must build up a case on their replies in ways similar to those Charles Darwin used in his study of the earthworm. Other equally important agencies in research are the insects, fungi, and other micro-organisms which attack the plant and the animal. These are Nature's censors for indicating bad farming. To-day the policy is to destroy these priceless agencies and to perpetuate the inefficient crops and animals they are doing their best to remove. To-morrow we shall regard them as Nature's professors of agriculture and as an essential factor in any rational system of farming. Another valuable method of testing our practice is to observe the effect of time on the variety. If it shows a tendency to run out, something is wrong. If it seems to be permanent, our methods are correct. The efficiency of the agriculture of the future will therefore be measured by the reduction in the number of plant breeders. A few only will be needed when soils become fertile and remain so.

Nature has provided in the forest an example which can be safely copied in transforming wastes into humus—the key to prosperity. This is the basis of the Indore Process. Mixed vegetable and animal wastes can be converted into humus by fungi and bacteria in ninety days, provided they are supplied with water, sufficient air, and a base for neutralizing excessive acidity. As the compost heap is alive, it needs just as much care and attention as the live stock on the farm; otherwise humus of the best quality will not be obtained.

The first step in the manufacture of humus, in countries like Great Britain, is to reform the manure heap—the weakest link in Western agriculture. It is biologically unbalanced because the micro-organisms are deprived of two things needed to make humus—cellulose and sufficient air. It is chemically unstable because it cannot hold itself together—valuable nitrogen and ammonia are being lost to the atmosphere. The urban centres can help agriculture, and incidentally themselves, by providing the farmers with pulverized town wastes for diluting their manure heaps and, by releasing, for agriculture and horticulture, the vast volumes of humus lying idle in the controlled tips.

The utilization of humus by the crop depends partly on the mycorrhizal association—the living fungous bridge which connects soil and sap. Nature has gone to great pains to perfect the work of the green leaf by the previous digestion of carbohydrates and proteins. We must make the fullest use of this machinery by keeping up the humus content of the soil. When this is done, quality and health appear in the crop and in the live stock.

Evidence is accumulating that such healthy produce is an important factor in the well-being of mankind. That our own health is not satisfactory is indicated by one example. Carrel states that in the United States alone no less than £700,000,000 a year is spent in medical care. This sum does not include the loss of efficiency resulting from illness. If the restitution of the manurial rights of the soils of the United States can avoid even a quarter of this heavy burden, its importance to the community and to the future of the American people needs no argument. The prophet is always at the mercy of events; nevertheless, I venture to conclude this book with the forecast that at least half the illnesses of mankind will disappear once our food supplies are raised from fertile soil and consumed in a fresh condition.

APPENDIXES

APPENDIX A

COMPOST MANUFACTURE ON A TEA ESTATE IN BENGAL

THE Gandrapara Tea Estate is situated about 5 miles south of the foot-hills of the Himalayas in North-East India in the district called the Dooars (the doors of Bhutan). The Estate covers 2,796 acres, of which 1,242 are under tea, and it includes 10 acres of tea seed bushes. There are also paddy or rice land, fuel reserve, thatch reserves, bamboos, tung oil, and grazing land. The rainfall varies from 85 to 160 inches and this amount falls between the middle of April and the middle of October, when it is hot and steamy and everything seems to grow.

The cold-weather months are delightful, but from March till the monsoon breaks in June the climate is very trying.

There are approximately 2,200 coolies housed on the garden; most of these originally came from Nagpur, but have been resident for a number of years. The Estate is a fairly healthy one, being on a plateau between large rivers; and there are no streams of any kind near or running through the property. All drainage is taken into a near-by forest and into waste land. The coolies are provided with houses, water-supply, firewood, medicines, and medical attention free, and when ill they are cared for in the hospital freely. Ante-natal and post-natal cases receive careful attention and are inspected by the European Medical Officer each week and paid a bonus; careful records of babies and their weights are kept and their feeding studied; the Company provides feeding-bottles, 'Cow and Gate' food, and other requirements to build up a coming healthy labour force. As we survey all living things on the earth to-day we have little cause to be proud of the use to which we have put our knowledge of the natural sciences. Soil, plant, animal, and man himself—are they not all ailing under our care?

The tea plant requires nutrition and Sir Albert Howard not only wants to increase the quality of human food, but in order that it may be of proper standard, he wants to improve the quality of plant food. That is to say, he considers the fundamental problem is the improve-ment of the soil itself—making it healthy and fertile. 'A fertile soil,' he says, 'rich in humus, needs nothing more in the way of manure: the crop requires no protection from pests: it looks after itself.' ...

In 1934 the manufacture of humus on a small scale was instituted under the 'Indore method' advocated by Sir Albert Howard. The humus is manufactured from the waste products of tea estates. All available vegetable matter of every description, such as *Ageratum*, weeds, thatch, leaves, and so forth, are carefully collected and stacked, put into pits in layers, sprinkled with urinated earth to which a handful of wood ashes has been added, then a layer of broken-up dung, and soiled bedding; the contents are then watered with a fine spray—not too much water but well moistened. This charging process is continued till the pit is full to a depth of from 3 to 4 feet, each layer being watered with a fine spray as before.

To do all this it was most necessary to have a central factory, so that the work could be controlled and the cost kept as low as possible. A central factory was erected; details are given in the plan (Plate X); there are 41 pits each $31 \times 15 \times 3$ feet deep; the roofs over these pits are 33 by 17 feet, space between sheds 12 feet and between lines of sheds 30 feet; also between sheds to fencing 30 feet; this allows materials to be carted direct to the pits and also leaves room for finished material. Water has been laid on, a 2-inch pipe with 1-inch standards and hydrants 54 feet apart, allowing the hose to reach all pits. A fine spreader-jet is used; rain-sprinklers are also employed with a fine spray. The communal cow-sheds are situated adjacent to the humus factory and are 50 by 15 feet each and can accommodate 200 head of cattle: the enclosure, 173 by 57 feet, is also used to provide outside sleeping accommodation; there is a water trough 11 feet 6 inches long by 3 feet wide to provide water for the animals at all times; the living houses of the cowherds are near to the site. An office, store, and chowkidar's house are in the factory enclosure. The main cart-road to the lines runs parallel with the enclosure and during the cold weather all traffic to and from the lines passes over this road, where material that requires to be broken down is laid and changed daily as required. Water for the factory has a good head and is plentiful, the main cock for the supply being controlled from the office on the site. All pits are numbered, and records of material used in each pit are kept, including cost; turning dates and costs, temperatures, watering and lifting, &c., are kept in detail; weighments are only taken when the humus is applied so as to ascertain tasks and tons per acre of application to mature tea, nurseries, tung *barees*, seed-bearing bushes, or weak plants.

The communal cow-sheds and enclosure are bedded with jungle and this is removed as required for the charging of the pits. I have tried out pits with brick vents, but I consider that a few hollow bamboos placed in the pits give a better aeration and these vents make it possible to increase the output per pit, as the fermenting mass can be made 4 to 5 feet deep. Much care has to be taken at the charging of the pits so that no trampling takes place and a large board across the pits saves the coolies from pressing down the material when charging. At the first turn all woody material that has not broken down by carts passing

PLATE X

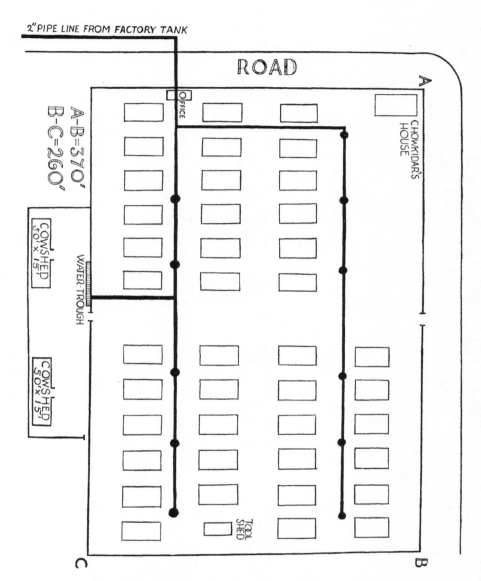

PLAN OF THE COMPOST FACTORY, GANDRAPARA TEA ESTATE

PLATE XI

COMPOSTING AT GANDRAPARA

A. Covered and uncovered pits. *B.* Roofing a pit.
C. Cutting *Ageratum*

over it is chopped up by a sharp hoe, thus ensuring that full fermentation may act, and fungous growth is general.

With the arrangement of the humus factory compost can be made at any time of the year, the normal process taking about three months. With the central factory much better supervision can be given, and a better class of humus is made. That made outside and alongside the raw material and left for the rains to break down acts quite well, but the finished product is not nearly so good. It therefore pays to cut and wither the material and transport it to the central factory as far as possible.

In the cold weather, a great deal of sheet-composting can be done. After pruning, the humus is spread at 5 tons to the acre; and hoed in with the prunings; the bulk of prunings varies, but on some sections up to 16 tons per acre have been hoed in with the humus and excellent results are being obtained.

On many gardens the supply of available cow-dung manure and green material is nothing like equal to the demand. Many agriculturists try to make up the shortage by such expedients as hoeing-in of green crops and the use of shade trees or any decaying vegetable material that may be obtainable; on practically all gardens some use is made of all forms of organic materials, and fertility is kept up by these means. It is significant to note that, for many years now, manufacturers who specialize in compound manures usually make a range of special fertilizers that contain an appreciable percentage of humus. The importance of supplying soils with the humus they need is obvious. I have not space to consider the important question of facilitating the work of the soil-bacteria, but it has to be acknowledged that a supply of available humus is essential to their well-being and beneficial activities.

Without the beneficial soil-bacteria there could be no growth, and it follows that, however correctly we may use chemical fertilizers according to some theoretical standard, if there is not in the soil a supply of available humus, there will be disappointing crops, weak bushes, blighted and diseased frames. Also it will be to the good if every means whereby humus can be supplied to the soil in a practical and economical way can receive the sympathetic attention of those who at the present time mould agricultural opinion.

To the above must be added the aeration of the soil by drainage and shade, and I am afraid that many planters and estates do not understand this most important operation in the cultivation of tea. To maintain the fertility we must have good drainage, shade trees, tillage of various descriptions, and manuring. Compost is essential, artificials are a tonic, while humus is a food and goes to capital account.

This has been most marked in the season just closing. From October 1938 to April 20th, 1939, there had been less than 1½ inches of rain, and consequently the gardens that suffered most from drought were those that had little store of organic material—drainage, feeding of the soil, and establishment of shade trees being at fault.

Coolies are allowed to keep their own animals, which graze free on the Company's land, and the following census gives an idea of what is on the property: 133 buffaloes, 115 bullocks, 612 cows, 466 calves, 21 ponies, 384 goats, 64 pigs: in all, 1,795 animals.

During the past two years practically no chemical manure or sprays for disease and pest-control have been used, the output for the past year of humus was about 3,085 tons, while a further 1,270 tons of forest leaf-mould have been applied. The cost of making and applying the former is Rs. 2/8/6 per ton, and cost and applying forest leaf-mould is Rs. 1/3/9 per ton.

The conversion of vegetable and animal waste into humus has been followed by a definite improvement in soil-fertility.

The return to the soil of all organic waste in a natural cycle is considered by many scientists to be the mode of obtaining the best-tasting tea, and to resisting pest and disease.

Nature's way, they claim, is still the best way.

GANDRAPARA TEA ESTATE,
BANARHAT, P.O.   J. C. WATSON.
18 *November*, 1939.

PLATE XII

A.

B.

C.

COMPOSTING AT GANDRAPARA

A. Communal cowshed. *B.* Crushing woody material by road
traffic. *C.* Sheet-composting of tea prunings

COMPOST MAKING AT CHIPOLI, SOUTHERN RHODESIA

COMPOST of a kind has been made at Chipoli for a number of years, but till Sir Albert Howard's methods were mastered some years ago the waste of material had been considerable, the product unsatisfactory, and the cost, in comparison with that now produced, high.

Deep pits were used and the process was chiefly carried out under anaerobic conditions, with the result that it took many months and most of the nitrogen was lost. Farm-yard manure was stored either in the stockyards or in large solid heaps, with the result that when the mass was broken up to be carted on to the fields most of the nitrogen had been lost, and much of the coarse grass, reeds, and similar matter used for bedding remained fairly well preserved, much as bog oak in the mud, and the process of decomposition remained to be completed in the soil with the growing crop, much to the detriment of the latter.

At Chipoli the compost-field has been laid out on the same principle as at Indore. Water is laid on and standpipes are situated at regular intervals. One-inch rubber hose-pipes are used to spray the water on to the compost heaps. With this arrangement compost can be made at any time of the year, the normal process taking almost exactly three months.

It has been claimed that it is cheaper to make compost in heaps alongside the lands where the raw material is grown and to rely on the rains for the water. If the rains are regular this acts quite well; but this is not always the case, and with the interruption of the process the finished product is not so good. Another objection to making compost away from an artificial water-supply is that the material to be composted cannot be used the same season, with the result that a year is lost. I have known seasons when the sequence of the rainfall has been unsuitable for the completion of the manufacture, with the result that the position of a farmer who had been relying on this method for the provision of compost to maintain the fertility of his lands would be serious. The cost of a water-supply is a small insurance premium to pay for certainty of manufacture.

I find pits unnecessary even in the hot weather. If the heaps are sprayed over every day it is quite enough to maintain the correct degree of moisture, and one native can easily control 500 tons. To apply water in buckets is not satisfactory: the material does not get a uniform wetting—too much is thrown on one place and too little on another.

As the heaps are being turned, a controlled spray keeps the moisture-content correct.

The question of cost is raised against the central manufactory. I am of opinion that the small extra cost of transportation is far more than offset by better supervision and the control of the process. The cost of moving the raw material can be reduced by stacking the *san* hemp, or whatever is being used, in heaps, and allowing it to rot to a certain extent. This considerably reduces the bulk.

The material used for making compost at Chipoli is mostly coarse velt grass which is cut from river banks, dongas, and wherever it is available; next in bulk is *san* hemp grown for the purpose, and then rushes, crop wastes, weeds, garden refuse, and so forth.

Compost is returned to the *san* hemp stubble and the land then ploughed. In the past large quantities of *san* hemp have been ploughed into the land to maintain the humus supply. In some seasons this works quite well, but in others, owing to unfavourable weather conditions, quantities of unrotted vegetable matter are left on or under the surface, to be decomposed the following season before a crop can be planted. By cutting and composting this surface growth and returning it to the land, everything is ready for planting as soon as the rains commence. Again, compost made from combined animal and vegetable waste has evidently some great advantage over humus derived from the top growth of a green crop only.

In making the heaps, a layer of vegetable waste is put down; the heaps are built about 25 yards long and 15 feet wide. Dung and urine-saturated bedding is then laid on top and on this is spread the correct quantity of soil and wood-ash; the whole is then wetted from the hose-pipe and the process repeated till the heap is some 3 feet high. Heating commences at once, and after some ten days, when fungous growth has become general, the heap is turned and more water applied if required. Two heaps are made side by side and if the bulk has become reduced considerably, as generally happens by the time the third turning is due, the two heaps are thrown into one. This maintains the bulk and so ensures that the process goes on properly without any interruption. Should action appear slow at the first turning, compost from another heap—which is being turned for the second time and in which action has been normal—is scattered among the material as it is being turned; inoculation thus takes place and the process starts up as it should.

I have found that a mixture of grass and *san* hemp acts much better than either *san* hemp or grass alone.

It has been the practice to lay coarse material on the roads and to allow wagon traffic to pass over it for some time; this breaks it down and action is much more satisfactory when manufacture commences. A better plan is to pass all the raw material through the stock-yards, where it becomes impregnated with urine and dung and gets broken up at the same time by being trampled. All that is then necessary in making the heaps is to mix this material with soil and wood-ash and moisten it.

It has always been the routine to broadcast some form of phosphatic fertilizer on the lands. This is now added direct to the compost heaps and so reaches the fields when the compost is being spread. The cheapest form of manure to be bought locally is bone-meal; besides phosphate this contains about 4 per cent. of nitrogen. Dried grass, the chief source of raw material, contains about one-half per cent. of nitrogen; this is very low, so that the addition of the extra nitrogen in the bone-meal assists the manufacture, and none of the nitrogen is lost. This addition of bone-meal is simply a local variation and is in no way necessary for the working of the process.

This year a spell of very wet weather converted the open cattle-yards into a quagmire. As soon as possible the sodden bedding and manure was carted on to the compost field and built into heaps with a liberal interbedding of soil. The material was so sodden that it packed tightly and a dark-coloured liquid exuded from the heaps. The material was turned immediately and more soil added, which absorbed the free liquid. After an interval of three days, a further turning took place and with this the swarms of flies, which had followed the manure from the yards, disappeared. Heating was slow, so a further turn was given; at each turn the heaps became more porous; with this last turn, heating became rapid and the fungous growth started, normal compost manufacture having commenced. Now that the principle of turning with the consequent aeration is understood, losses which took place in the past through improper storage will be avoided. One is reminded of the family midden in countries like Belgium, with their offensive smells and clouds of flies; if the composting principle was understood, what loss could be avoided and how much more sanitary would conditions become!

The chief inquiry with many people before commencing compost-making is that of cost. This largely depends on local conditions. Labour costs and the ease with which the raw materials can be collected are the chief factors. I happen to grow tobacco and to use wood as fuel for curing; my tobacco barns are close to the compost field so that my supply of wood-ash is both plentiful and handy. The stock-yards are situated close at hand, through which in future it is hoped to pass all the vegetable wastes. It has been found that *san* hemp hay makes an excellent stock food; stacks of this will be made alongside the compost field and feeding pens put up where the working oxen can get a daily ration, the refuse being put on to the heaps.

Compost making has been going on for too short a time here to be able to give definite costs. A particular operation that costs a certain sum this year may have its cost halved next year as methods of working are improved. As an approximate indication, however, the following will serve.

On a basis of turning out 1,000 tons of finished compost, collecting all the
raw material and spreading the compost on the field.

For those who do not know, a South African wagon is generally
18 feet long; it is drawn by a span of sixteen oxen and holds a normal
load of 5 tons. For carting vegetable wastes I make a framework of
gum poles which sits on the top of the wagon and so greatly increases
its carrying power for bulky materials. Sometimes two or three such
wagons work on compost making the same day, and sometimes not
any, but an average would be one wagon full time for four months.
Such a wagon requires a driver and a leader and two other men for
loading and unloading with the help of the driver and sometimes the
leader. Labour for cutting and collecting the coarse grass, reeds, &c.,
works out at about ten natives every day for two months. The *san*
hemp is cut with a mowing-machine and collected with a sweep, say
four natives for one month. As regards the manufacture itself, four
natives for five months can attend to everything. This gives a total of
1,800 native days. For spreading the finished compost some people
use a manure-spreader, which does an excellent job, but such an implement would be too slow for us.

On Chipoli three wagons are used for spreading at the same time;
each wagon carries something over 3 tons of finished compost. Four
natives fill the wagons at the heaps and as soon as a wagon arrives in
the field it is boarded by four other natives with shovels or forks who
spread the compost on a strip of predetermined width as the wagon
moves slowly along. On an average, taking adjacent and more remote
lands, one wagon makes eight trips a day; thus with a total of fourteen
natives we spread some 75 tons of compost a day. Spreading 1,000 tons
thus takes 200 native days. In other words, the whole operation from
cutting the waste materials to spreading the finished product on the
land takes 2,000 native days. This means that the work of two natives
for one day is required for each ton of compost made and spread on
the land.

To the above must, of course, be added the upkeep and depreciation
on wagons, mowing-machine, &c., when engaged on this work, but
this is quite a small item. The ox is not taken into account, as not only
does he assist in the manufacture by providing waste material, but when
his term of service has come to an end he is fattened up and sold to
the butcher, generally for a sum at least as much as he cost.

My water service, made from material purchased from an old mine,
was written off after the first season.

I made the statement recently before the Natural Resources Commission that, if compost making became general in Southern Rhodesia,
the agricultural output of the country could be doubled without any
more new land being brought under cultivation.

Last year the bill for artificials on Chipoli was roughly half of what

PLATE XIII

A.

B.

COMPOST-MAKING AT CHIPOLI, SOUTHERN RHODESIA

A. General view of composting area.   *B.* Watering the heaps

it used to be, and if the state of the growing crops is any indication the out-turn will increase by fifty per cent.

This season compost has been used on citrus, maize, tobacco, monkey nuts, and potatoes. A neighbour was persuaded, somewhat against his will, to make some compost; this he did and applied it to land on which he planted tobacco: he now tells me that that particular tobacco was much the best on his farm.

Some photographs published in the *Rhodesia Agricultural Journal* show in a striking manner the drought-resisting properties imparted to land after being dressed with compost. The maize plants on the land to which compost had been added show almost no signs of distress, while those alongside on land that had no compost are all shrivelled up. Properly made compost has the property of fixing a certain amount of atmospheric nitrogen. To do this to the best advantage it appears necessary that the manufacture should be carried out as quickly as possible. There must be no interruption, and the material must on no account be allowed to dry out or to become too wet. I am inclined to use more soil than is absolutely necessary; it costs nothing, and the small extra charge in transport is more than covered by its presence as a form of insurance against any nitrogen that might be given off, which it tends to grasp and fix. We at the present stage know little about mycorrhiza, but it is probable that an excess of soil is not a disadvantage where this is concerned.

Where the acreage is large and the compost will not go round it all, it is probably better to give a medium dressing to a larger acreage than a heavy dressing to a smaller one. A dressing of about 5 tons to the acre is about the minimum for ordinary crops, but for such things as potatoes and truck crops at least 10 tons to the acre should be given, and, if available, considerably more. It must be borne in mind that much of the soil in Rhodesia has been so depleted of humus that in order to bring it properly to life again much heavier dressings of compost will be necessary now than when it has once attained natural conditions.

The more I see of compost-making the more necessary it appears to be to let the material have continuous access to air. This, as has been previously explained, can be done by frequent turning, and if turned quickly very little heat need be lost and no interruption in the process takes place.

An advance to better air-supply would be a series of brick flues under the heaps. But under my conditions, with the position of the heaps continually changing and with Scotch carts and heavy wagons continually moving among the heaps, the flues would always be getting broken. Six-inch pipes with slots cut in the sides and the piece of metal from the slot hinged on one side and turned outwards on each side so as to form an air-space in the compost, with a continual supply of air from inside the pipe, would probably act quite well, the advantage being that such pipes would be portable and would be laid down just before the heap was about to be built. The disadvantage is, of course, one of cost. To go even farther, a small oil-driven compressor, such as is used to drive a pneumatic hammer, and mounted on a wheelbarrow or small hand-truck and connected to a pipe by rubber hose, could be used. The pipe would be, say, 1 inch in diameter and pointed at one end. For a distance of perhaps 18 inches from the point small holes would be drilled. The pipe would be pushed into the heap at the centre and air pumped in, the operation being repeated at perhaps distances of 3 feet. A large number of heaps could be treated with forced aeration in a day, and if this method resulted in the fixation of only a few extra pounds of nitrogen per ton it might be well worth while.

This is, however, perhaps going too far at the present stage. The great beauty of Sir Albert Howard's method is its simplicity. It can be used in native villages by primitive people using their own tools equally well as on the most up-to-date estates using elaborate machinery.

I am glad to say that the Rhodesian Government have laid it down that compost-making is to be taught at all native agricultural instructional centres. Interest in the matter will gradually awaken. I have already had old men from neighbouring villages come in to see how manure was made from dry grass.

We are on the eve of the compost era. Had its principles been applied years ago, the desolation that has taken place in the Middle Western States of America could have been avoided. The so-called 'law of diminishing returns' is seen to apply only to those who do not really understand the soil and treat it as Nature meant it to be treated. Rhodesia is fortunately a young country and the destruction of its soil has not gone very far, comparatively speaking.

If compost making becomes general, which means thorough rebuilding of the soil and so providing it with greater fertility, greater power to withstand droughts through its enhanced ability to absorb the rainfall, much of which now runs needlessly to the ocean, a great change in the agricultural outlook will take place. The present system, employed by many, of mining instead of farming the soil, of stimulating it to the last extent with artificials, and—when it has been killed—abandoning it, must be exchanged for real soil-building according to Nature's methods. Only in this way can disaster, examples of which can be seen all round, be avoided, and the land be made to produce what it was intended to produce before our too clever methods were employed upon it.

J. M. MOUBRAY.

CHIPOLI, SHAMVA,
SOUTHERN RHODESIA.

2 *February* 1939.

THE MANUFACTURE OF HUMUS FROM THE WASTES OF THE TOWN AND THE VILLAGE[1]

By SIR ALBERT HOWARD, C.I.E., M.A.,

Formerly Director of the Institute of Plant Industry, Indore, Central India, and Agricultural Adviser to States in Central India and Rajputana.

THE forest suggests the basic principle underlying the correct disposal of town and village wastes in the tropics. The residues of the trees and of the animal life, met with in all woodlands, become mixed on the floor of the forest, and are converted into humus through the agency of fungi and bacteria. The process is sanitary throughout and there is no nuisance of any kind. Nature's method of dealing with forest wastes is to convert them into an essential manure for the trees by means of continuous oxidation. The manufacture of humus from agricultural and urban wastes by the Indore Process depends on the same principle —an adequate supply of oxygen throughout the conversion.

THE INDORE PROCESS

The Indore Process, originally devised for the manufacture of humus from the waste products of agriculture, has provided a simple solution for the sanitary disposal of night soil and town wastes. The method is a composting process. All interested in tropical hygiene will find a detailed account of the bio-chemical principles underlying the Indore Process, and of the practical working of the method, in the five papers cited at the end of this note.

HUMUS MANUFACTURE AT TOLLYGUNGE, CALCUTTA

Perhaps the best way of introducing the application of the Indore Process to town wastes will be to give an account of the recent work done by Mr. E. F. Watson, O.B.E., at the Tollygunge Municipality trenching ground, near Calcutta.

The conversion of house refuse and night soil into humus is carried out in brick-lined pits, 2 feet deep, the edges of which are protected by a brick kerb.[2] Each compartment of the pit has a capacity of 500 cubic feet and channels for aeration and drainage are made in the

---

[1] Reprinted from a paper read at the Health Congress at the Royal Sanitary Institute held at Portsmouth from July 11th to 16th, 1938.

[2] The guard rim is made of two bricks laid flat in cement mortar, two quarter-inch rods in the join serving as reinforcement. The upper brick should project 1 inch over the pit to form a lip for preventing the escape of fly larvae.

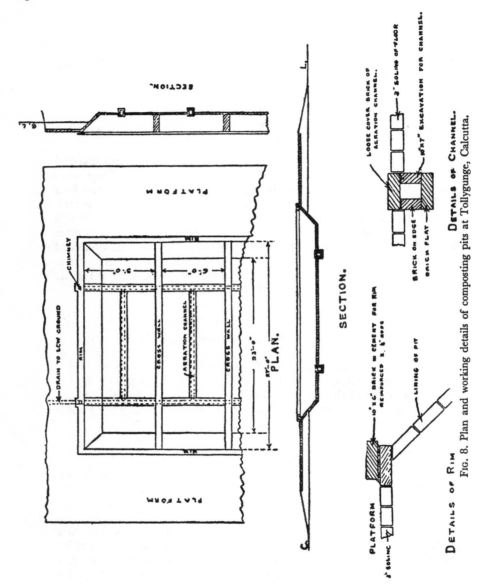

DETAILS OF RIM.

SECTION.

DETAILS OF CHANNEL.

FIG. 8. Plan and working details of composting pits at Tollygunge, Calcutta.

floor.[1] The area round the pit is protected by brick soling. Working details of these composting pits are shown in Fig. 1.

The method of charging the pits is most important, as success depends on correct procedure at this point. To begin with, a cartload of unsorted refuse is tipped into the pit from the charging platform and spread by drag rakes (Fig. 9) to make a layer 3 or 4 inches thick. Another cartload of refuse is then tipped on this layer and raked into a slope reaching from the edge to the middle of the pit and occupying its whole width. The surface of this slope is slightly hollowed by raking a little refuse from the centre to the sides. A little refuse is also raked on to the sill at the road edge to receive any night soil spilt on it by tipping. Half a cartload of night soil is then tipped on the slope and with the moistened refuse below it is drawn by drag-rakes in small lots until the breadth of the pit is covered. This done, the remaining half-load of night soil is poured on the freshly exposed surface of the slope and distribution by raking repeated until the slope (and the refuse on the sill) is altogether removed and forms a layer over the whole of the pit being charged. Another cartload of refuse is then tipped, another slope made, the sill covered, night soil added and raked away. The whole group of operations is so repeated until the pit is charged. This takes 2 days. The top layer of the first day's charge must be covered with 2 inches of refuse and left unmixed with the layer below. This helps to keep uniform moisture and heat in the mixed charge and to prevent the access of flies. The last operation on the second day is to make a vacant space at the end of each pit for subsequent turning and also for assisting drainage after heavy rains. This is done by drawing up 2 feet of the contents at one end over the rest. The surface is then raked level and covered with a thin layer of dry house refuse.

There is no odour from a pit properly filled, because the copious aeration effectively suppresses all nuisance. Smell, therefore, can be made use of in the practical control of the work; if there is any nuisance the staff employed is not doing the charging properly. They are either leaving pockets of night soil or else definite layers of this material, both of which interfere with aeration and so produce smell.

First Turn.—Five days from the start the contents of the pit must be turned. The object of this turn is to complete the mixing and to turn into the middle, and so destroy the fly larvae which have been forced to the cooler surfaces by the heat of the mass.

The original mixing of the heap, as well as the turning, are best done

[1] The aeration channels are covered with bricks laid open-jointed, and are carried up at the ends into chimneys open to the wind. By this means air permeates the fermenting mass from below. At one point these channels are continued as a drain to the nearest low-lying land. It is an advantage when bricking the pits to give a slight slope towards the aeration channels as this helps in keeping the pits dry in wet weather.

with long-handled manure drags by men standing on the division walls or on a rough plank spanning them.

Second Turn. After a further ten days the mass is turned a second time, by which time all trace of night soil will have disappeared.

Watering. In dry weather it may be necessary to sprinkle a little water on the refuse at each turn. The contents must be kept damp *but not wet.*

In very wet weather, when the surface of the pits is kept continually cool by rain, there is much development of fly larvae before the first

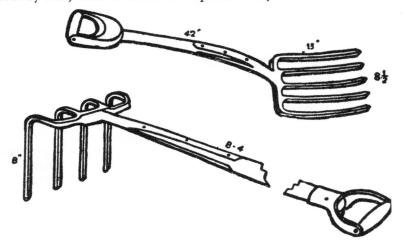

Fig. 9. Long-handled drag-rake and fork used in composting.

turn, but since these cannot escape and are turned into the hot mass and destroyed before they can emerge as flies, no nuisance results. Flies, therefore, are most useful in providing another means of automatic control.

Ripening of the Compost. After a further two weeks the material is removed from the pits to the platform for ripening. The whole process, therefore, takes one month. The stacks of ripening compost should be 4 feet high, arranged clear of the loading platform on a stacking ground running between two lines of pits (Fig. 10). The stacking process permits of sorting. Any material not sufficiently broken down, such as sticks, leather, coco-nut husks, and tin cans are picked out and thrown into an adjacent pit for further treatment. Inert materials such as brickbats and potsherds are thrown on the roads for metalling. Hand-picking is easy at this stage, as the contents of the pit have been converted into a rough, inoffensive compost. The ripening process is completed in one month, when the humus can be used either for manuring vacant land or as a top dressing for growing crops.

Cost. The capital cost is very small. A population of 5,000 in India yields some 250 cubic feet of house refuse daily, enough to mix with all the night soil. This will require a compost factory of sixteen pits of 500 cubic feet each, one pit being filled in two days (Fig. 10). With roads, platforms, and tools this costs from Rs. 1,000 to Rs. 1,500. The daily output is 150 cubic feet of finished compost, which finds a ready

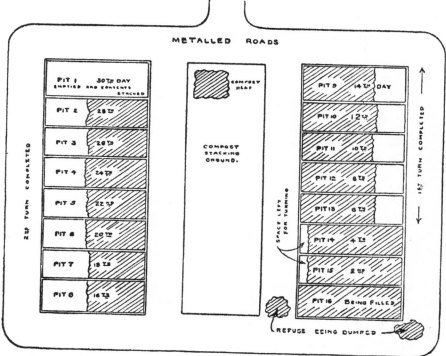

FIG. 10. Plan of compost factory at Tollygunge, Calcutta, after one month's use.

sale at Rs. 5 to Rs. 7. At the lower figure the sale proceeds of the first year will be about Rs. 1,800. This more than covers the working expenses. A factory of this size will need a permanent staff of five men.

A SIMPLE INSTALLATION FOR A VILLAGE

When a rural community is too poor to own conservancy carts or to construct brick-lined pits, composting can be carried out in an open trench on any high ground, without the use of partition walls. A trench and the whole sequence of operations is shown in Fig. 11 (p. 241).

The difficulty with unlined pits is the escape of fly larvae which breed in the walls of the trench and in the stacks of ripening compost. This disadvantage can be overcome either by bricking the vertical walls or by keeping fowls, which thrive on the larvae.

SOME FURTHER DEVELOPMENTS

The Use of Humus in Collecting Night Soil. There is one weak point in these two applications of the Indore Process to urban wastes. In both cases night soil is collected, transported, and composted in the crude state. This gives time for putrefaction to begin and for nuisance to develop. It can be prevented by the use of humus in the latrine pails, which ensures the oxidation of the night soil from the moment of deposition, and so prevents nuisance and the breeding of flies. The pails should contain at least 3 inches of dry humus when brought into use each day, and the droppings should be covered with a similar layer of humus when the pails are emptied into the conservancy carts. In this way putrefaction and smell will be avoided; the composting process will start in the pails themselves. The use of humus will augment the volume and weight of the night soil handled, but this increase in the work will be offset by the greater efficiency of composting, by the suppression of smell and flies, and by a considerable reduction in the loss of combined nitrogen.

Composting Night Soil and Town Wastes in Small Pits. Night soil can be composted in small pits without the labour of turning. These pits can be of any convenient size, such as 2 feet by $1\frac{1}{2}$ feet and 9 inches deep, and can be dug in lines (separated by a foot of undisturbed soil) in any area devoted to vegetables or crops. Into the floor of the pits a fork is driven deeply and worked from side to side to aerate the subsoil and to provide for drainage after heavy rain. The pits are then one-third filled with town or vegetable waste, or a mixture of both, and then covered with a thin layer of night soil and compost from the latrine pails. The pit is then nearly filled with more waste, after which the pit is topped up with a 3-inch layer of loose soil. The pit now becomes a small composting chamber, in which the wastes and night soil are rapidly converted into humus without any more attention. After three or four months the pits will be full of finished compost and alive with earthworms. A mixed crop of maize and some pulse like the pigeon pea (*Cajanus indicus*) can then be sown on the rows of pits as the rainfall permits, and gradually earthed up with the surplus soil. The maize will ripen first, leaving the land in pigeon pea. The next year the pits can be repeated in the vacant spaces between the lines of pulse. In two seasons soil fit for vegetables can be prepared.

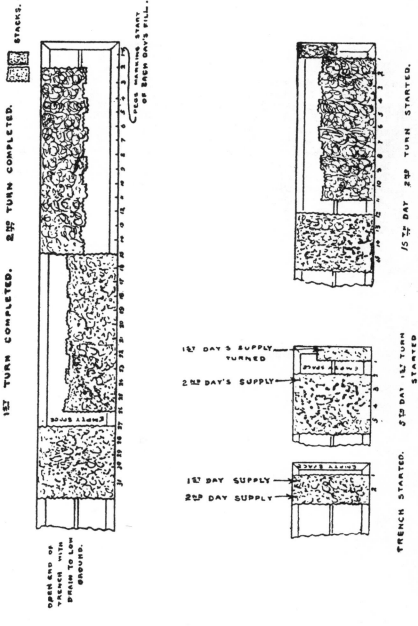

FIG. 11. Plan of a simple composting trench for a village. First and second day's produce stacked and Section 2 ready for refilling

LITERATURE

HOWARD, A., and WAD, Y. D. *The Waste Products of Agriculture: Their Utilization as Humus*. Oxford University Press, 1931.

JACKSON, F. K., and WAD, Y. D. 'The Sanitary Disposal and Agricultural Utilization of Habitation Wastes by the Indore Method', *Indian Medical Gazette*, lxix, February 1934.

HOWARD, A. 'The Manufacture of Humus by the Indore Method', *Journal of the Royal Society of Arts*, November 22nd, 1935, and December 18th, 1936. (These papers have been reprinted in pamphlet form and copies can be obtained from The Secretary, Royal Society of Arts, John Street, Adelphi, W.C. 2.)

WATSON, E. F. 'A Boon to Smaller Municipalities: The Disposal of House Refuse and Night Soil by the Indore Method', *The Commercial and Technical Journal*, Calcutta, October 1936. (This paper is now out of print, but the substance has been incorporated in a lecture by Sir Albert Howard to the Ross Institute of Tropical Hygiene on June 17th, 1937. Copies can be obtained on application to the lecturer at 14 Liskeard Gardens, Blackheath, S.E. 3.

HOWARD, A. 'Soil Fertility, Nutrition and Health', *Chemistry and Industry*, vol. lvi, no. 52, December 25th, 1937.

INDEX

Aberystwyth, 189.

acidity,
  necessity for neutralizing in humus manufacture, 44.
  due to use of sulphate of ammonia, 59.
  of irrigation water from sisal wastes, 77.

Adco process, 51, 52.

aeration, need for, in conversion of town wastes, 112, 115; in humus manufacture, 233, 237.

aeration trenches, effect on young trees under grass, 128-9; effect on superficial root-system, 129.

aerobic phase, need for water and air, 48; in humus production from green crop, 93.

aerotropism of roots, 121, 125, 134.

afforestation, need for to combat soil erosion, 147.

agriculture,
  various systems of, 1.
  ancient systems, 6-9.
  peasant systems of the Orient, 9-17.
  methods of the West, 17-20.
  most successful man-made systems, 25.
  balance of, 25.
  stabilization of systems, 32.
  disturbance of balance, 33.
  disintegration in the West, 35.
  necessity to human existence, 104.
  need for cultivation of new types of crops, 184.
  basic purpose of, 198.
  need for co-operation between science and the farmer, 221.

agricultural chemistry, 181; decline of, 182.

agricultural economics, 186; fallacy of, 197-8.

agricultural research,
  need for, 40.
  obsolete character of present-day organization, 41.
  growth of progress, 181.
  need for practical agricultural knowledge, 182.
  organization in Great Britain, 187-8.
  gap between science and practice, 189.
  in the Empire, 191.
  evils of team work, 193-5.
  insistence of quantitative results, 196.
  misuse of, 198-9.
  vital need for new system, 221-2.

Agricultural Research Council, 187, 188.

Agricultural Research Institute, 160.

agricultural sciences,
  growth of, 186.

agricultural sciences (cont.)
  classifications, 187.
  small value of results, 190.
  in British agriculture, 195-6.

agronomy, 195.

algae, 33.
  importance in tropical soil, 34.
  oxygen supply dependent on, 81.
  immobilization of nitrate by, 88, 91.
  attraction by old tree bark, 126.

alkali lands,
  formation of, 148.
  poisoning of Siberian Lakes by, 149.
  stages of development, 150.
  appearance of salts on soil, 151.
  occurrence of, 152.
  effect on crops, 152-3.
  theory of reclamation, 153-4.
  irrigation as a cause of, 154-5.

All-India Board of Agriculture, 201.

alluvium of NW. India, 71.

Amani Research Institute, 191.

ammonia, given off by humus, 27; from plant decomposition, 89.

ammonium sulphate, 69.

anaerobic conditions, induced too soon, 45; establishment of in compost pits, 48, 49.

anaerobic fermentation, 111; in formation of alkali tracts, 151.

anaerobic organisms, 50.

animal wastes, 36.
  use by Chinese for composting, 14.
  loss to Western agriculture, 37.
  available, 42.
  substitutes for, 42.
  importance of collection, 43.
  necessity for in humus manufacture, 65.
  need for in sisal waste conversion, 78.
  in nitrate immobilization, 92.
  compared to human wastes, 112.

apple, as a mycorrhiza-former, 167.

Application of Science to Crop-Production, note, 41.

arable pans, 137.

artificial manures,
  extensive use in Western agriculture, 18.
  spread of disease through, 19.
  folly of, 38.
  failure in sugar-cane estates, 67-8.
  effect on vegetable crops, 82.
  effect on cereal crops, 136.
  formation of alkali-tracts due to, 151.
  connexion with plant and animal diseases, 167.

artificial manures (*cont.*)
  growth of popularity, 185.
  multiplicity of results of, 185.
  growth of disease, 186.
  poisoning of soil by, 220.
Assam Valley, 57.

bacteria, 23, 35.
  use in humus manufacture, 45.
  activity in compost-pits, 49.
  in sisal waste conversion, 76.
  action on organic matter in refuse tips, 109–10.
bail system, 99.
bananas,
  research at Trinidad, 191.
  breeding of new variety, 192.
  failure of research, 193.
  need for soil aeration and humus, 194.
Banerji, S. C., 201.
Barber, Dr., 184, 204.
barley, 12, 13, 184.
basic slag, 100.
beans, 13, 94, 107.
Beavan, —, 184.
Belcher, Major, 54, 55.
Belgian Congo, 55.
Bihar, 202.
Biologists in Search of Material, 178.
biology, applied, in manurial problems of tea, 61; problem of in green-manuring, 95.
black alkali, 151.
Bodiam, 92–3, 106, 107.
boll-weevils, 165.
boll-worms, 165.
Borbhetta, 58.
botanical research, 183–4.
Boussingault, 181.
British Association, 83.
British Medical Journal, vii, 113, 178.
Broadbalk, 29, 182, 220.
Bromley-Davenport, Sir William, 178.
Butler, —, 75.

cacao,
  research at Trinidad, 191–2.
  failure of research, 193.
  need for humus and soil aeration, 194.
Cachar, 57.
Cambridge research laboratories, 156.
carbohydrates, 24.
  synthesis in sugar-cane, 68, 214.
  synthesis in cotton, 74.
carbon dioxide, 24.
  given off by humus, 27.
  produced in refuse tips, 110.
  in pore spaces under grass, 130.
  action on rocks, 139.
carbon monoxide, 111.
Carlyle, Sir Robert, 161.
Carrel, Alexis, 222, 224.

catch crops, 33, 35, 92.
cattle,
  at Indore, 50.
  wastes in Western agriculture, 50.
  on sugar estates, 66, 67.
  effect of humus on health of, 100.
  diseases of, 162.
  eradication of disease, 161–2.
cellulose, 35.
  impure, 36.
  bark as a protection of, 42.
  breaking down during synthesis of humus, 49.
  content in green crops, 89.
Central Indian States, 40.
Ceylon Tea Plantations Co., vii.
Chhatari, Nawab of, 215.
chickweed, 92, 93.
China, 32; simple system of agriculture, 51.
Chintamani, Sir Chirravoori, 215.
Chipoli,
  manufacture of humus at, 229–36.
  materials used at, 230.
  cost of composting at, 231–2.
chlorophyll, importance of, 23; lack of in soil organisms, 24.
Clarke, George, vi, 88, 200, 201, 202, 204, 206, 215, 217.
clover, 13, 107.
  synthesis of mycorrhiza by, 98.
  requirements compared to grass, 99.
  development of new variety, 184.
  red clover, 97.
cluster bean, 97.
codlin moth, 167.
coffee, 12, 53–6.
  effect of Indore Process, 54.
  trials of Indore Process, 166.
Coimbatore, 13, 184, 217.
cold weather crops, 35.
Coleyana Estate, 73.
composting,
  objection of lack of material, 42.
  collection of animal residue, 43.
  dependence on work of fungi and bacteria, 51.
  conversion of tea prunings and green-manure, 63–4.
  difficulty of utilizing cane trash, 68–70.
  best method of utilizing humus, 70.
  adopted for cotton cultivation, 73.
  materials from green crops, 95.
  of human wastes, 112.
  of night soil, 113, 240.
  to check disease transference, 168.
  early method at Chipoli, 229.
compost heaps,
  use of animal residues, 36.
  need for water and air, 45.
  value compared with pits, 46.
  need for protection from heavy rain, 62.
  layout for sisal waste conversion, 76.

compost heaps (*cont.*)
 need for close attention, 223.
 addition of phosphatic fertilizer at Chipoli, 231.
compost pits,
 advantage over heaps, 45.
 necessity for careful drainage, 46.
 size and method of charging, 46.
 turning material, 48.
 need for protection from heavy rain, 62.
 use on tea estate, 64.
Condesuyu, 6.
Constitution and Functions of the Agricultural Research Council, 186.
Conway, —, 6.
Cook, O. F., 5, 6.
Cornell, 131.
Correns, —, 183.
Corve Dale, 99.
Costa Rica, 56.
cotton, 11, 12.
 colloids in cultivation, 33, 71.
 limiting factor in cultivation, 71.
 need for good farming methods, 72.
 results with Indore Process, 72.
 preparation of humus from wastes, 72–3.
 Indore Process tried in Africa, 73.
 reason for reaction with humus, 74.
 low nitrogen content of stalks, 95.
 reaction to alkali, 153.
 disease-free crops at Indore, 164.
 trials of Indore Process, 166.
 development of new varieties, 184.
Coventry, Bernard, 159.
cover crops, 140, 147.
crop production,
 finance of, 38.
 limitation of at Pusa, 118.
 limitation of chemistry, 183.
 plants as chief agents, 183.
 practical results of research, 184.
crop yield, 39.
crops,
 Nature's use of parasites, 156.
 methods of dealing with disease, 157–8.
 first investigation of crop pests, 159.
 indirect methods of combating parasites, 160.
 basic reasons of crop disease, 161.
 disease in deep root varieties, 162.
 vital necessity of farm-yard manure, 165.
Crotolarias, 55.
cultivation, importance of, 23–4; effect on insect pests, 163, 164.
custard apple, 120, 122, 125, 126, 127, 128, 133.

Darjeeling, 57.
Darwin, Charles, 183, 189, 222.

decay, in the life cycle, 25; in the production of humus, 27.
Dehra Dun, 56.
Der Tropenpflanzer, 53.
Development Commission, 187.
diet, effect of natural soil products, 177.
Dooars, 57, 62, 225.
drainage, necessary for soil fertility, 63; to combat root disease, 65.
dredge corn, 13.
Drought Investigation Commission, 142.
dub grass, 97.
Duncan, Walter, & Co., vii, 57.
dunder, 66, 69.
dung,
 use as fuel in India, 14.
 as an essential in humus synthesis, 69.
 compared to humus as fertilizer, 107.
dust bowls of North America, 42.
Dutch lights, cultivation under, 13.
Dymond, G. C., 69, 70.

earthworms,
 drainage and aeration channels of, 2.
 role in soil aeration, 136.
 need for in breaking up of arable pans, 137.
 Darwin's account of work, 183.
East African Agricultural Journal, 75.
elephant grass, 55.
Empire Cotton Growing Association, 73.
Empire Cotton Growing Review, 74, 75.
En Busca del Humus, 56.
entomology, 40.
Esquimaux, 177.
Evans, Sir Geoffrey, 191, 195.

factory wastes, 33.
Farmer's Club, 101, 107.
Farmer's Weekly, vii, 86.
farmyard manure,
 lack of in Western agriculture, 35.
 effect on green crop, 88.
 need for by lucerne crop, 97.
 need for by *san* hemp crop, 97.
 in cultivation of *san* hemp, 164.
 a basis for plant and animal health, 165.
 problem of sufficient supply, 165.
Farnham Royal, 157.
Farrer, —, 184.
fermentation,
 during humus manufacture, 45.
 slowing down by rain, 46.
 retarded by lack of water, 48.
 during composting, 49.
 difficulty in use of cane trash, 69.
 necessity for 'starter' with sisal waste, 76.
 Pasteur's work on, 183.
field records, 34.

field trials, 27.
filter cake, 69.
Finlay, James & Co., vii, 57.
fodder, soil fertility and quality compared, 29.
food preservation, 19.
foodstuffs, necessity for new system of grading, 221.
food values, effect on human health, 172, 175; future research at Peckham, 178.
foot-and-mouth disease, 19, 162.
forest trees,
    Nature's method of manuring, 2.
    effect of grass in India, 132–6.
    compared to fruit trees, 134.
fruit trees, compared to forest trees, 134; pests at Quetta, 163.
fungi,
    as aid to synthesis of humus, 35, 42, 48, 76.
    as pointer to unsuitable methods of cultivation, 160.
    role in controlling unsuitable varieties of crops, 161.
    development of in sugar-cane, 206.
fungous bridge, vi, 99.
fungous diseases,
    on sugar estates, 67.
    in cotton crops, 74.
    in vine cultivation, 85.

Gandrapara Tea Estate, 62, 64.
    details of, 225–8.
    output of humus at, 228.
garden cities, need for new planning, 114–15.
Gatooma, 73.
Gilgit Agency, agricultural system in, 173.
Gorakhpur, 202.
gram, 12.
    crop attacked by insects, 164.
    cultivation in sugar-cane trenches, 213.
grape gardens, lack of disease in Quetta valley, 163.
grass,
    essential requirements of, 97.
    requirements compared to clover, 99.
    competition with trees, 117.
    injurious effect on fruit trees, 119, 125–30.
    effect on root system, 126–7.
    disappearance of soil nitrates, 131.
    effect on Indian forest trees, 132–6.
    role in prevention of soil erosion, 146, 147.
    necessity for in surface drains, 147.
    growth on reclaimed alkali soil, 154.
    development of new varieties, 184.
grass-land management,
    underlying principles in Great Britain, 96.

grass-land management (cont.)
    role of fungous bridge, 99.
    bail system, 99.
    use of basic slag, 100.
    partial and complete cultivation, 101.
    world-wide problem, 102.
    improvements necessary, 102–3.
Great Rift Valley, 54.
green crop, tea composition at various stages, 89; value in humus production, 95.
green-fly,
    infection of crops by, 162.
    fruit trees affected by, 163.
    method of checking, 164.
    infection of apple trees by, 167.
green-manuring,
    general failure of, 87.
    chief factors of, 88.
    for increase of soil nutrients, 89–90.
    failure in tropical agriculture, 91.
    uses in agriculture, 91–5.
    reform of, 95.
greensand, 136.
Greenwell, Sir Bernard, 101, 107, 179.
Grenada, 134.
Grogan, Major, 54, 55, 75.
guava,
    examination of root system, 124.
    benefit of irrigation, 125.
    effect of grass, 125, 126, 132.
    effect of aeration trenches, 129, 134.
    prolific growth of, 135.
Guinness, Messrs. Arthur, Son & Co., vii, 92, 106.
Gwalior State, 145; Maharaja of, 146.
gypsum, 153.
gyrotiller, 18.

Hamilton, Mrs. V. M., viii.
Harler, Dr., 56, 57, 58.
Haynes, L. P., 92, 106.
health of mankind,
    related to fertile soil produce, 171–2, 177–9.
    research work necessary, 179–80.
    dependence on improvement in agriculture, 224.
Heitland, —, 7.
Hilgard, E. W., 151, 183.
Hiralal, —, 123.
Holbeach, 139.
hops, 92–3.
    hop gardens, 106.
    experiments on, 159.
Hosier, A. J., 99, 102, 103.
household refuse, 33.
humus,
    description and importance of, 24 et seq.
    effect on crops, 28.
    as a key to agriculture, 30.
    replenishment by animals, 34.

humus (*cont.*)
sources of in Western agriculture, 35.
conversion of vegetable wastes, 36.
gap between crop usage and artificial replacement, 37.
necessity for adequate supply in plant breeding, 39.
need for, to compensate higher crop yields, 40.
discovery of practical method of manufacture, 41.
necessity for animal waste products in make-up, 42.
low value of chemical activated type, 43.
neutralizing of acidity in manufacture, 44.
necessity for water and air in composting, 45.
methods of converting wastes, 45.
period of manufacture, 50.
output and storage of, 50.
moisture content of, 51.
manufacture from tea estate wastes, 57, 225–7.
in tea cultivation, 57–8, 61–2.
as opposed to chemical fertilizers, 59.
necessity for healthy plant development, 61.
compared to sulphate of ammonia, 62.
as mycorrhiza former in sugar-cane, 68.
yield from cane trash, 70.
conversion of sisal waste, 75.
effect on maize crops, 79.
control of witch-weed by, 79.
increase in rice crops, 81.
response of transplanted crop, 81.
results of intensive use, 84.
full benefit of drainage, 85.
yield from green crop decay, 89–91.
production from green-manure crop, 93–4.
need for by *san* hemp crop, 97.
need by grass-lands, 99–103.
need for cellulose and lignin in synthesis, 106.
costs compared to artificials, 107.
formation in refuse tips, 109–12.
effect of carbon dioxide in soil on, 130.
natural conversion of leaves, 135.
formation under turf, 137.
prevention of soil erosion by, 146.
need for in alkali lands, 150.
need for in *san* hemp cultivation, 164.
importance in combating plant disease, 167.
restoration of soil fertility, 167.
effect of crops on live stock, 179.
Liebig theory, 181–2.
need for in cacao and banana cultivation, 194.

humus (*cont.*)
for sugar-cane cultivation, 212.
time required for formation, 212.
importance of supply to soils, 227.
assistance in resisting drought, 233.
use in Rhodesia, 233.
use at Tollygunge, 235–9.
in collection of night soil, 240.
Hunzas, effect of diet on physique, 173; effect of rock powder on agriculture, 174.
Hussain, Sheikh Mohomed Naib, 201.
Hutchinson, H. B., 51.
hydrogen, 111.

Iceni Nurseries, 47; results with Indore Process, 83–4.
Imperial Bureaux of Entomology and Mycology, 157.
Imperial College of Tropical Agriculture, Trinidad, 191; failure of research at, 193.
India, 34, 68, 70; important cotton areas, 71.
Indian Central Cotton Committee, 40, 165.
Indian Tea Association, 57.
indigo, 12.
Indore,
method of keeping cattle at, 43.
methods of controlling cattle disease at, 162.
lack of infectious cattle disease at, 165.
Indore City, 112.
Indore Institute, foundation of, 166.
Indore Process, 39, 52.
principles underlying process, 41.
method of manufacture, 41 *et seq.*
use of in Western and tropical agriculture, 44.
laboratory investigations, 51.
publication of first results, 53.
success in coffee cultivation, 55.
spread of in America, 56.
applied to tea, 57.
early trials, 59.
results with cotton crop, 72–3.
for sisal waste conversion, 76.
benefit to maize crop, 78.
results with rice crop, 80.
benefit to vine cultivation, 86.
first trials with human wastes, 112.
evolution of, 166.
provision of manure for Indian villages, 215, 217.
Indore Residency, 112.
Insch, James, 57.
insects, indicators of unsuitable methods of cultivation, 160–1.
insect diseases,
on sugar estates, 67.
in cotton crops, 74.

insect diseases (*cont.*)
in vine cultivation, 85.
Institute of Plant Industry, 39, 73.
foundation of, 40.
note, 41.
output of humus at, 50, 164.
cotton growing survey by, 71.
testing of Indore Process at, 166.
iodine, 96.
irrigation,
staircase farming in Peru, 5.
at La Crau, 29.
for cotton cultivation, 71, 73.
by sisal waste drainage, 76.
in humus production from green crop,
93.
low oxygen content of water, 116.
effect on root systems, 121.
additional benefits to guava, 125.
in Nira valley experiment, 149.
formation of alkali tracts by, 150.
basin system, 154.
cause of alkali lands, 154–5.
effect on insect pests, 163, 164.
in sugar cultivation, 206, 217.

Jackson, F. K., 112.
Jenkins, W. J., 73.
Johannsen, —, 183.
Jones, B. B., and Owen, F., 109.
Jouques, —, 85.
Journal of the Ministry of Agriculture, 53.
Journal of the Royal Society of Arts, vii, 53,
191.
jute, 11, 12; development of new varieties,
184.

Kanan Devan Hills Produce Company,
57.
Kenya, 55, 56.
humus in tea cultivation, 58.
exhaustion of maize soils, 78.
soil erosion in, 142.
Kerr, Mrs., 80.
Kerr, Rev. G. M., 80.
Kilvert, William, 99.
King, *Farmers of Forty Centuries*, 10, 51.
King, F. H., 154, 183.
Kingatori Estate, 53, 54, 55.
Kipping, Professor, 200.

La Crau, 98, 99; outstanding fertility of,
29.
Land Settlement Association, 113.
Lathyrus sativus, 162.
Layzell, Major S. C., 75, 76.
leaching, 34.
of nitrates from humus, 50.
loss of nitrates by, 88, 92.
Levisohn, Dr. Ida, vi, 60, 82, 98, 99, 123,
166.

leys, 33, 35, 147.
short, 96.
temporary, 136, 137, 140.
lichen, 24; attraction by old tree bark, 126.
Liebig, J., 181–2, 220.
Liebig tradition, 37, 182, 185; conception
of soil fertility, 183.
lignin, content in green crop, 89; protection against fungi, 42.
lime, irrigation water neutralized by, 77.
lime sulphur wash, 84.
lime trees, 122, 125, 126, 128.
litchi, 124, 125, 126, 127, 128, 131, 132.
locusts, role in soil erosion, 142.
loess soils,
sub-soiling of, 137.
erosion of, 145.
alkali patches in, 149.
study of, 163.
maintenance of porosity, 164.
loquat, 124, 125, 126, 127, 128, 131, 132.
Louisiana, 68.
Lowdermilk, —, 144.
lucerne, 97, 137, 164; for curing alkali in
soil, 153.
lupins, 87.

McCarrison, Sir Robert, 172, 173, 174, 176.
maize, 12, 13.
foodstuff for British dairy animals, 78.
exhaustion of crop soils, 78.
synthesis of humus from wastes, 79.
need for new system of grading and
marketing, 79.
stalks used for cattle bedding, 95.
need for organic matter in soil, 97.
trials of Indore Process, 166.
protection from witch-weed, 166.
malaria, diminution of mosquitoes, 65.
malnutrition in sugar-cane, 68.
Malwa Bhil Corps, 112.
mango, 122, 124, 125, 126, 127, 128.
Mann, Dr. H. H., 55.
manure,
substitute for soil fertility, 33.
green, 33, 35, 63, 69.
farming without, 34.
chemical, 37.
uselessness of artificials, 37.
use in compost pits, 47.
artificial assessment of value, 66.
pen, 66, 67, 68.
kraal, 69.
lack of for market gardening, 82.
for sugar cultivation, 205, 208, 210–14.
Manures and Manuring, note, 105.
manuring,
tea, 62.
contrasting methods in sugar cultivation, 67.
necessity for, for fresh food crops, 177.
fragmentation of research, 196.

Marden Park, 101, 107, 179.
mares' tails, 137.
Masefield, G. H., vii.
Medical Testament, 175, 176, 177, 178, 179, 180.
Meerut, 202.
megass, 66.
Mendel's law, 183.
methane, 110.
micro-organisms,
  activity during humus manufacture, 45, 49.
  breakdown of cellulose by, 49.
  Waksman's theory, 51.
  breaking down of cane trash by, 68.
  need for oxygen, 85.
  effect of decomposition on, 89.
  in green crop decay, 90, 91.
  formation of nitrates from soil, 91.
  synthesis of humus, 100, 106.
millet, 12, 13, 97.
Ministry of Agriculture, 187, 188.
Mississippi, erosion by, 145.
mixed crops, 1, 13; to assist aeration, 116.
mixed farming, 1, 32.
molasses, 69.
Mommsen, Theodor, 7, 8.
monoculture, 1, 4, 13, 17.
monsoon, 32, 62, 93.
  effect on root systems, 121.
  rain on guava root system, 124.
  use of aeration trenches, 129.
  formation of pans as a result of, 137.
Montealegre, Señor Don Mariano, 56.
Morford, Kenneth, 61, 62.
Moubray, Captain J. M., vii, 79.
Mount Vernon, 61, 64.
Mukerjee, Jatrindra Nath, 131.
mustard, 92.
mycelium,
  in tea plants, 60.
  in sugar-cane, 68, 206.
  in meadow herbage, 98.
  in surface root systems, 122.
  action in young roots, 123.
mycology, 40; Panama disease in bananas, 192.
mycorrhiza,
  in sugar-cane, 68.
  importance to cotton, 74, 75.
  rice as a builder of, 82.
  vines as a builder of, 86.
  synthesis by tropical legumes, 98.
  synthesis by grass family, 99.
  synthesis by fruit trees, 122.
  synthesis by apple, 167.
  synthesis by sugar-cane, 206.
mycorrhizal association, vi, 23, 24, 25.
  essential link in plant nutrition, 37.
  in coffee crop, 56.
  in tea crop, 60, 61, 66.
  lack of in starved soil, 60.

mycorrhizal association (cont.)
  necessity for good crops, 62.
  in sugar-cane roots, 68, 206.
  in Egyptian cotton crops, 75.
  in rice cultivation, 81, 82.
  in vines, 86.
  in tropical leguminous plants, 98.
  role in promoting growth, 100.
  in fruit trees, 122.
  role in plant nutrition, 122.
  effect of carbon dioxide, 130.
  factor in health of crop, 166.
  in roots of banana and cacao, 193, 194.
  establishment of necessary conditions in sugar cultivation, 212.
  role of humus in, 223.

Nairobi, 53, 54, 55.
Natal, 68, 69, 70.
National Geographical Society, 5.
National Health Insurance Act, 175.
Nature, 75.
Nicholson, —, 86.
night soil, 37.
  for conversion into humus, 112, 237.
  use of humus in collection of, 240.
nitrate immobilization, 92, 94; requirements of sugar-cane, 209.
nitrates, absence of in heavy soils, 100.
nitrification,
  difficulty of in rice cultivation, 81.
  effect of carbon dioxide in soil, 130.
  of organic matter in soil, 183.
  in green-manure, 212.
nitrogen,
  need for conservation in soil, 15, 16.
  content in soil, 27, 35.
  immobilization of, 34.
  fixation of, 34.
  in animal and industrial residues, 35, 36.
  in tea cultivation, 59.
  content in sugar-cane humus, 70.
  content in sisal waste humus, 76.
  in green-manuring, 88.
  loss of in green crop decay, 91.
  produced in refuse tips, 110, 111.
  content in fungous tissue, 166.
  avoidance of loss in sugar-cane production, 212.
nitrogen cycle, 208, 214.
North Bihar, fertility of soils, 135; alkali lands at, 148, 149.
Nottingham University College, 200.
NPK fertilizers, 18, 66, 82, 87; rise of importance, 182, 185.
Nullatanni, 57.
nutrition,
  value of various foods, 171–3.
  illness through misuse, 175.
  results of specialized diets, 176.
  importance of food quality, 178.
  research work necessary, 179–80.

Nuwara Eliya, 56.
Nyasaland, use of humus in tea cultivation, 58.

oats, 107.
O'Brien, G. W., 66.
oil seeds, 12.
opium, 12.
organic manures, value of, 111; use in sugar cultivation, 205.
organic matter,
  in the soil, 34.
  loss of fertility through non-replacement, 62.
  in black cotton soils, 71.
  response of rice to, 81.
  natural formation of nitrates from, 91.
  percentage in household refuse, 110.
  gases produced by decay, 111.
  non-renewal by failure of suitable rotation, 137.
osmotic pressure of solutions, 152.
Ossendowski, F., 149.
Oudh, 216.
oxidation processes, 116; in Indore Process, 235.
oxygen,
  need of by heavy soils, 100.
  produced in refuse tips, 110.
  lack of in swamp rice cultivation, 111.
  necessity for by soil organisms, 116.
  exhaustion of in alkali soils, 151.

Panama disease in bananas, 192.
parasites,
  reason for activities, 39.
  in crops, 156.
  in insect control, 157.
  lack of in Indian peasant agriculture, 160.
  use in growing healthy crops, 161.
  need for research reform, 169.
Pasteur, Louis, 183, 189.
Pathans, 173.
pathology, 40; vegetable, 158.
peach trees, 120, 126, 127.
Pearse, Dr., 178.
peas as a rotation crop, 94.
peasant farming, 9–17.
Peat, J. E., 73.
Peckham Health Centre, 178.
pedology, 183.
Peru, 5, 152.
phosphates, use by trees, 3; search for by tree roots, 135.
phosphoric acid, 70, 111.
phosphorus, 3, 36.
physiological diseases of plants, 156.
Picton, Dr. L. J., vii, 113.
pigeon pea, 13, 15, 97.
  use as a sub-soiler, 137.
  for curing alkali in soils, 153.

pipul, 148.
plant-breeding, 40; basic necessities for success, 73.
plough-pans, effect on soils, 136.
plum trees,
  study of root system, 120.
  effect of grass on roots, 127.
  effect of aeration trenches, 128.
Poore, Dr. G. V., 113–14.
Pope, Professor, 200.
pore-space, 23, 24.
potash, 36, 111; used by trees, 3, 135.
potassium sulphate, 69.
potato, eelworm in crops, 195; cultivation compared to sugar-cane, 204.
potato disease, in British crops, 195; methods for combating, 196.
Prescott, J. A., 6.
Proceedings of the Royal Society of London, vii, 117.
proteins, 24.
  in sugar-cane, 68.
  synthesis in cotton, 74.
  in green crop, 91.
Provincial Advisory Service, 187.
Provincial Agricultural Dept., 73.
pulses, 12.
Punjab, 202.
Pusa Agricultural Research Institute, 39, 40, 72, 117, 118, 121, 124, 127, 133, 159, 162, 165.

quarantine, to check importation of crop parasites, 158.
Quetta Experimental Station, 153, 162; lack of infectious cattle disease at, 165.
Quetta Valley, 149, 153, 163.

rainfall,
  Nature's method of conservation, 2.
  in humus manufacture, 45.
  effect on humus synthesis from green-manure crop, 93.
rats, diet experiments on, 173; value of experiments, 174.
Rayner, Dr. M. C., vi, 60, 75, 166.
reafforestation, 32.
Rees, Sir Milsom, 55.
Report of the Imperial Agricultural Research Conference, note, 191.
Report on Agricultural Research in Great Britain, 186.
Research Institutes, 187, 188; lack of connexion with agriculture, 189.
Revista del Instituto de Defensa del Café de Costa Rica, vii, 53, 56.
Rhodesia, 55, 95.
  cotton growing in, 73.
  exhaustion of maize soils, 78, 166.
  growth of composting in, 229–34.
Rhodesia Agricultural Journal, 79, 233.

Rhodesia Herald, 73.
rice, 12.
  importance of in East, 15–16.
  cultivation in western India, 32.
  first trials of Indore Process, 80.
  effect of humus on, 81–2.
  effect on human physique, 173.
  development of new species, 184.
Richards, E. H., 51.
rinderpest, 162.
river Nen, 140.
  Ouse, 140.
  Welland, 140.
Rohilkhand, 202, 203.
Roosevelt, Franklin D., 141.
roots,
  importance of systems, 23.
  source of soil organic matter, 33.
  effect of humus on tea plant, 60–2.
root development,
  under influence of grass, 119.
  in Indian forest trees, 133.
  activity dependent on rainfall, 134.
root disease, treatment for soil fertility,
  63, 65.
root systems,
  effect of oxygen in soil, 117.
  study of plum tree, 120.
  effect of moisture on superficial system,
  121.
  effect of rainfall, 121, 124.
  effect of grass on, 126.
  effect of extra aeration, 129.
Rosa Sugar Factory, 204.
Rothamsted Experimental Station, 14,
  29, 34, 96, 168, 182.
Rowett Institute, 96.
Royal Botanic Gardens, 193.
Royal Sanitary Institute, vii, 112.
Royal Society of Arts, vii, 56.
Rural Hygiene, 113.
rye, 13.
rye-grass, 13, 97.

Sabet, Younis, 75.
St. Vincent, 135.
Sakrand, 73.
san hemp, 95, 97.
  crop attacked by mildew, 164.
  for humus manufacture, 230.
Sarda Canal, 218.
Saunders, —, 184.
Schloesing's method, 209.
School of Technology, Manchester,
  200.
Schultz-Lupitz, —, 87.
science, analytical methods of, 22.
seaweed, 33; for humus manufacture,
  36.
septicaemia, 162.
sereh, 204.
sewage systems, water-borne, 3;

shade trees, 63.
Shahjahanpur Experiment Station, 12,
  88, 94, 147, 201, 202, 204, 212, 213,
  214, 215, 218.
Shahjahanpur principle, 70.
shoddy, 36.
Sikhs, 177.
sisal,
  method of waste disposal, 75.
  conversion to humus, 76.
  drainage water used for irrigation, 76.
  need for intensive cultivation, 77.
  necessary condition for success in
  composting, 77–8.
  trials of Indore Process, 166.
Sisiphus, 178.
sodium carbonate, 151, 152.
sodium chloride, 151.
sodium sulphate, 151.
soil,
  fungi, 24.
  organic matter, 33.
  manurial rights of, 42.
  effect of poverty on green crops, 90.
  necessity for lucerne crop, 97.
  Nature's method of rotation, 139.
  Pasteur's work on bacteria, 183.
  as wealth of the world, 219.
soil aeration, 24, 70, 71, 72, 84, 85, 99.
  promoted by deep root systems, 15.
  effect of humus, 71.
  improved by cultivation, 101.
  allied factors, 117.
  relation to grass and trees, 118.
  reaction of tree roots, to, 121.
  effect of sub-soil water, 124.
  effect on root development, 128.
  effect of ground water, 135.
  restoration to arable pans, 137.
  lack of in alkali lands, 152.
  growth of plant disease, 162.
  connexion with insect attack, 162.
  effect on insect pests, 163.
  in cacao and banana cultivation, 194.
  in establishment of mycorrhiza, 207.
  necessity in tea cultivation, 227.
soil bacteriology, 183.
soil erosion, 59.
  prevention of, 63.
  in cotton areas, 74.
  in black cotton soils, 137, 145.
  related to infertility, 140.
  in the U.S.A., 140–1.
  in South Africa, 142.
  essential remedy, 143.
  control in Japan, 143–4.
  deforestation as chief cause, 143.
  control by forest trees, 144, 146.
  in China, 144–5.
  solution of the problem, 146.
  remedies necessary, 147.
  growth of menace, 219.

soil fertility,
  relation to commerce, 11.
  in Eastern peasant farming, 16.
  decline of in Western agriculture, 20.
  nature of, 22 *et seq.*
  effect on live stock, 30.
  restoration and use of, 33–8.
  need for in tea cultivation, 59.
  increased by use of humus, 60, 93.
  research work in cotton cultivation, 74.
  need for mycorrhizal association, 74, 167.
  in maize production, 78.
  early theories of, 87.
  necessity for in green-manuring, 91.
  need of grass-lands for, 102.
  testing of grass-lands, 102–3.
  influenced by root development, 135.
  necessity for to combat soil erosion 147.
  in reclamation of alkali tracts, 154.
  as basis of plant health, 165, 167.
  restored by use of humus, 167.
  relation to healthy people, 172.
  need for research, 180.
  necessity for safe-guarding, 219.
  loss of, through use of artificial manures, 220.
soil gases, carbon dioxide content, 130.
soil management, Nature's method, 1; disease caused by inefficiency, 171.
soil physics, 183.
Soils and Men, 142.
Southwark, pulverized town wastes from, 106; increasing demand for wastes, 108.
sour limes, 124.
Stapledon, Sir George, 102.
Steel, Octavius, & Co., vii.
Steiner, Rudolf, v, 14.
Striga lutea (witch-weed), 79.
  as indicator of soil fertility, 79.
  protection of maize by humus from, 166.
sub-soiling,
  to assist aeration, 101, 116, 132.
  importance in aeration in East, 137.
  Oriental method of breaking up pans, 137.
  natural action of Nature, 138.
sub-soil drainage, 116; experiment in Nira valley, 149.
Suez Canal, 11.
sugar-cane, 12, 13.
  wastes available, 66.
  decline in quality, 67.
  results from natural and chemical fertilizers, 67.
  need for conversion of wastes to humus, 68, 69.
  heavy crops from specially prepared soil, 94.

sugar-cane (*cont.*)
  low nitrogen content, 95.
  maximum yields obtained, 97.
  as a mycorrhiza former, 98.
  trials of Indore Process, 166.
  new varieties of, 184.
  successful research by Clarke, 200.
  success of research in India, 200–18.
  cultivation in India, 202.
  need for nitrates in soil, 210.
  success of green-manuring, 210–14.
sulphate of ammonia,
  compared to humus, 27, 62.
  used in tea cultivation, 58–9.
  effect on fertile soil, 62.
  substitute for nitrogen in guava cultivation, 131.
  formation of alkali tracts due to, 151.
sulphuretted hydrogen, 111.
  contamination of wells in alkali zones, 149.
  from bacilli in Siberian lakes, 149.
  in formation of alkali soils, 151.
sun as an aid to growth, 23.
surface drainage, necessity in sugar-cane cultivation, 207.
Sylhet, 57.
Szira-Kul, Lake, 149.

tamarind, 148.
Tambe, G. C., 70.
Tanganyika, 55, 56.
Taveta, 55, 75, 77.
tea, 12.
  conversion of wastes into humus, 57.
  methods of manuring, 58.
  improvement with compost manures, 60.
  comparison between humus and artificial manures, 61–2.
  need for more seed gardens, 65.
  effect of artificial manures, 66.
  trials of Indore Process, 166.
Tenney, F. G., 89.
Tephrosias, 55.
Terai, 57.
termites, use of tunnels by tree roots, 133; attacks on sugar-cane, 206.
The Waste Products of Agriculture, v, 41, 54, 80.
The Wheel of Health, 174.
thrips, 192.
tilth, maintenance of, 28; restoration of, 33.
Timiriasev Academy, Moscow, 146.
Timson, —, 79.
tobacco, 12.
  cure of virus diseased crop, 163.
  trials of Indore Process, 166.
  growing at Chipoli, 231.
Tocklai Research Station, 57, 58, 59.
Tokyo Forestry Board, 144.

Tollygunge, humus manufacture at, 235-9.
town wastes,
    loss to agriculture, 104.
    reform needed, 105.
    use of pulverized wastes, 106-7.
    potential value of amount available, 108.
    controlled tipping, 109-12.
    for providing humus of the future, 223.
    composting in small pits, 240.
Transactions of the British Mycological Society, 74.
Travancore, 56, 57, 58.
trees, for maintenance of soil fertility, 136.
Tristan da Cunha, 176; diet of islanders, 177.
Tukoganj, 123.

Uganda, 55.
Ultor glacier, 6.
United Provinces, 34.
urine, 37.
    use by Chinese for composting, 14.
    substitutes for, 36.
    use in Indore Process, 43.
    use of impregnated soil, 44, 47.
    essential unit in humus manufacture, 51, 69.
usar plains of North India, 150.

vegetables,
    wastes, 37, 41-2, 46, 63.
    need for supply of humus, 82.
    effect of chemical fertilizers, 82.
ventilation of the soil, 24.
vetch, 13.
vine, 85-6.
virus diseases of plants, 156.
    problem of protecting bananas from, 193.
    in potato crop, 195.

Wad, Y. D., 70, 75, 82, 112, 123.
Waksman, S. A., vii, 26, 27, 51, 88, 89.
Waldemar tea estate, 98.
Wallace, T., 57.
Walton, *Compleat Angler*, 30.
water, culture, 1; need for in humus manufacture, 45, 229.
water-hyacinth, 63, 65.
water-weeds, 33.
    for maintaining soil fertility, 36, 63.
    cultivation beside streams, 37.
watering, of compost pits, 47; during turning of material in pits, 49.
Watson, E. F., 235.
Watson, J. C., 62.
Welsh Plant Breeding Station, 96, 101.
West India Committee Circular, 56.
wheat, 12, 13, 107.
    continuous cropping in North America, 17.
    mechanization of cultivation, 18.
    as diet of northern India, 172.
    new varieties of, 184.
    cultivation in sugar-cane trenches, 213.
Whitmore, J. E. A. Wolryche, 78.
Williams, —, 146.
Williamson, Dr., 178.
Wilson, Captain, 83, 84, 85.
Winogradsky, —, 183.
witch-brown disease, 192.
witch-weed, 79, 166.
Woburn Experimental Station, 119, 125, 136.
Wrench, G. T., 174.
Wye College, 159.
Wythenshawe, 109.

Yellow River, soil erosion by, 145.